PREDICTION OF PERFORMANCE AND POLLUTANT EMISSION FROM PULVERIZED COAL UTILITY BOILERS

PREDICTION OF PERFORMANCE AND POLLUTANT EMISSION FROM PULVERIZED COAL UTILITY BOILERS

N. SPITZ
R. SAVELIEV
E. KORYTNI
M. PERELMAN
E. BAR-ZIV
AND
B. CHUDNOVSKY

Nova Science Publishers, Inc.
New York

Copyright © 2009 by Nova Science Publishers, Inc.

All rights reserved. No part of this book may be reproduced, stored in a retrieval system or transmitted in any form or by any means: electronic, electrostatic, magnetic, tape, mechanical photocopying, recording or otherwise without the written permission of the Publisher.

For permission to use material from this book please contact us:
Telephone 631-231-7269; Fax 631-231-8175
Web Site: http://www.novapublishers.com

NOTICE TO THE READER

The Publisher has taken reasonable care in the preparation of this book, but makes no expressed or implied warranty of any kind and assumes no responsibility for any errors or omissions. No liability is assumed for incidental or consequential damages in connection with or arising out of information contained in this book. The Publisher shall not be liable for any special, consequential, or exemplary damages resulting, in whole or in part, from the readers' use of, or reliance upon, this material.

Independent verification should be sought for any data, advice or recommendations contained in this book. In addition, no responsibility is assumed by the publisher for any injury and/or damage to persons or property arising from any methods, products, instructions, ideas or otherwise contained in this publication.

This publication is designed to provide accurate and authoritative information with regard to the subject matter covered herein. It is sold with the clear understanding that the Publisher is not engaged in rendering legal or any other professional services. If legal or any other expert assistance is required, the services of a competent person should be sought. FROM A DECLARATION OF PARTICIPANTS JOINTLY ADOPTED BY A COMMITTEE OF THE AMERICAN BAR ASSOCIATION AND A COMMITTEE OF PUBLISHERS.

LIBRARY OF CONGRESS CATALOGING-IN-PUBLICATION DATA

ISBN: 978-1-60741-184-0

Available upon request

Published by Nova Science Publishers, Inc. New York

CONTENTS

Preface		vii
Chapter 1	Introduction	1
Chapter 2	Methodology	3
Chapter 3	Prediction Results for Coals Not Tested by IEC	49
Chapter 4	Conclusions	61
References		63
Index		67

PREFACE

A three-step methodology was developed to provide reliable prediction of a coal's behavior in a utility boiler: (1) Extracting the combustion kinetic model parameters by combining experimental data from a pilot-scale test facility, Computational Fluid Dynamic (CFD) codes and an artificial neural network. While the combustion kinetic parameters used in the model code will not correspond to the combustion rate of a single particle of coal, these parameters do describe the combustion behavior of a "macroscopic" sample of tested coal. (2) Validation of the combustion kinetic model parameters by comparing diverse experimental data with simulation results calculated with the same set of model parameters. (3) The model parameters are then used for simulations of full-scale boilers using the same CFD code. For operational engineering information needed by the utility operator, we apply the predicted results to EXPERT SYSTEM, a boiler supervision system developed by Israel Electric Corporation (IEC). Four different bituminous and sub-bituminous coals with known behavior in IEC 550MW opposite-wall and 575MW tangential-fired boilers were used to show the adequacy of the methodology. The predictions are done with the CFD code, GLACIER, propriety of Reaction Engineering International (REI). Preconfigured GLACIER models of the test and full-scale furnaces were purchased from REI and validated by our group. This book chapter will include a detailed description of the methodology, test furnace facility and an example of the experimental and predictive combustion results from the four coals used to test the methodology. In addition, two previously unknown coals will be examined prior to their firing in the utility boilers and prediction of their behavior and operational parameters in the two boilers will be carried out.

Chapter 1

INTRODUCTION

Pulverized coal is an important fuel for electricity production [1] and will continue to be important for decades. Since coal is a natural resource that depends on many factors and parameters, it has variable properties and composition. Because of this heterogeneity, the combustion behavior and pollutant emissions are different for each coal. The days when utility companies used coal from the same mine for years are gone [2-3] and they are faced now with the challenge of firing very different types of coal in the same boiler. The great variability of the properties and composition of the different coals imposes tremendous operational difficulties and requires innovative approaches to aid in decision making and operational strategies. In addition to this, more stringent environmental regulations are being enforced [4] and the utility company needs to lower its emissions while staying profitable. Different types of coals combined with modifications to the combustion process can be a cost-effective way of improving combustion behavior and yet lowering pollutant emissions from power plants. Because of the variability of combustion behavior of each coal, utility companies go to great lengths to test the coals before purchasing them for use in their utility boiler. These tests, which include preliminary tests at coal site and then full-scale tests at operator's site, are very costly. Although much advanced in recent years, the predictions of Computational Fluid Dynamic (CFD) models for coal combustion are not sufficient alone to select a coal for full-scale use [1,5]. CFD models can give good predictions of coal combustion in utility boilers if the coal combustion kinetic model parameters are known. These kinetic parameters are usually determined from sub-models.

The aim of this study was to develop a low-cost method to predict the combustion behavior and pollutant emission from coals previously unknown to the utility company. To predict the combustion behavior and pollutant emissions of coal in pulverized-coal utility boilers, we developed a three-step methodology to provide reliable prediction of a coal's behavior in a utility boiler:

1. Extracting the combustion kinetic model parameters by combining experimental data from a pilot-scale test facility, CFD codes and an artificial neural network. While the combustion kinetic parameters used in the model code will not correspond to the combustion rate of a single particle of coal, these parameters do describe the combustion behavior of a "macroscopic" sample of tested coal.
2. Validation of the combustion kinetic model parameters by comparing diverse experimental data with simulation results calculated with the same set of model parameters.
3. The model parameters are then used for simulations of full-scale boilers using the same CFD code. For operational engineering information needed by the utility operator, we apply the predicted results to EXPERT SYSTEM, a boiler supervision system.

The predictions and full-scale tests were done on two boiler types: 550MW opposite-wall and 575MW tangential-fired. The coals tested and presented here are: (1) Billiton-BB Prime – a South African bituminous coal (Billiton or SA in text), (2) Glencore-Adaro – an Indonesian sub-bituminous coal (Adaro or Ad), (3) Drummond-La Loma – a Colombian bituminous coal (Drummond or Dr), for tangential-fired boiler only, (4) Glencore-Russian – a Russian bituminous coal (Russian or Gln). For both boilers, we predicted the behavior and emissions from two coals previously unknown to IEC: Guasare-Venezuelan (Venezuelan or Ven) – a Venezuelan bituminous coal and KPC-Melawan – an Indonesian sub-bituminous coal (Mel). For opposite-wall boiler we also simulated the combustion of Glencore-Russian coal. The predictions are done with the CFD code, GLACIER, propriety of Reaction Engineering International (REI). Preconfigured GLACIER models of the test and full-scale furnaces were purchased from REI and validated by our group.

Chapter 2

METHODOLOGY

The combustion and pollutant formation processes involved in utility boilers depend on (1) the geometry and materials of the combustion chamber that includes the burners, air inlets, coal and air feed factors; all of these are known by the utility operator, and (2) the composition and properties of the coal. Of the latter, the chemical composition and heat values can be analyzed in the chemist's laboratory but the CFD code includes twelve model parameters depicting kinetic and thermal properties of the coal; all of which cannot be analyzed or precisely estimated in the laboratory. If all coal quality parameters (thermodynamic, thermal, and optical properties as well as chemical composition and kinetic parameters) are known, one can use a sophisticated CFD code to simulate the performance of the system. However, only some of the required information is available and hence one must carry out, prior to the simulation, experimental testing of the coal in order to acquire the missing information required for the simulation. One can obtain all parameters required by carrying out testing in a series of laboratory-scale equipment. These measurements, however, may lack credibility in the representation of the coal/blend involved as very small samples (normally milligrams) are used in these laboratory testing. In order to obtain credible parameters, one needs to use a representative quantity of coal/blend, normally in the vicinity of 400-600 kg. This amount can be tested only in a pilot scale facility.

Coal combustion processes in test facilities, such as this one, are simplified because many of the physical and chemical processes that take place in a utility boiler are absent. Therefore, it is easier to validate the CFD combustion model for the test furnace and apply the validated model, with the same kinetic parameters but with different boundary conditions, to a utility boiler. The approach described here is based on a simplified kinetics model, shown in

Figure 1. The model includes a two-step devolatilization process with three parameters each (activation energy, pre-exponential factor, and weight fraction for each step), totaling 6 parameters for devolatilization; a char combustion model with two parameters (activation energy and pre-exponential factor); one parameter to represent the nitrogen distribution between char and volatile matter; one parameter to represent the distribution of HCN and NH_3, and one parameter to represent NO_x release from char. While it cannot be said that the kinetic parameters used in the model will correspond to the combustion rate of a single particle of coal, these parameters do describe the combustion behavior of a "macroscopic" sample of the tested coal. For the goal of predicting combustion behavior of coals in a utility boiler furnace, this simplified model used with the CFD code gave good results.

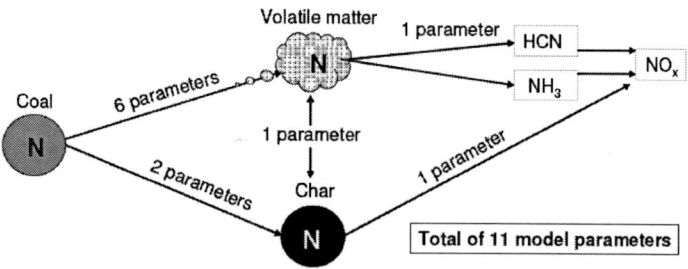

Figure 1. Kinetic model for devolatilization – combustion of coal.

A flow chart of the three-step methodology we developed is detailed in figure 2. From the flow chart one can see that there is no attempt to predict the combustion behavior of the utility boiler based on the combustion behavior of the test furnace. The three-steps are:

1. Obtaining the combustion kinetic model parameters required for the numerical simulations of utility boilers from a series of experiments in a 50 kW pilot-scale test facility, CFD codes and optimization algorithm. A pilot-scale furnace was constructed at our facility for this purpose. The predictions are done with the CFD code, GLACIER, propriety of Reaction Engineering International (REI). The optimization algorithm developed is based on an artificial neural network (ANN). To simplify interpretation of the combustion experiments the furnace was designed with axial symmetry and with a two-zone configuration: a well-mixed reaction zone followed by a plug flow region. Furthermore, the coal is burned within the test furnace at temperatures and concentrations

similar to those prevailing in a pulverized-coal furnace. The model parameters used in the simulation are modified until good agreement is obtained between the results of the numerical simulation and the experimental data from the test furnace. The drawback of the multi-parameter fitting technique is whether the set obtained is unique to the system or there are other sets that would also provide good agreements, due to numerical inter-compensation.

2. Validation of the combustion kinetic model parameters by comparison of different experimental data with simulation results obtained by the set of combustion kinetic parameters. To attain confidence in the fitting technique numerous experiments were carried out at different operating conditions (i.e.: stoichiometric ratios, coal size distributions, and staged burning). The same set of model parameters was used to check the correlation of the experimental and the numerical results. The more data to fit the more confidence there is in the obtained set of parameters. Yet, some uncertainty will always remain, though very difficult to quantify.

3. The extracted combustion kinetic model parameters are then used for simulations of full-scale boilers using the same CFD code. For operational engineering information needed by the utility operator, we apply the predicted results to EXPERT SYSTEM, a boiler supervision system developed by Israel Electric Corporation (IEC).

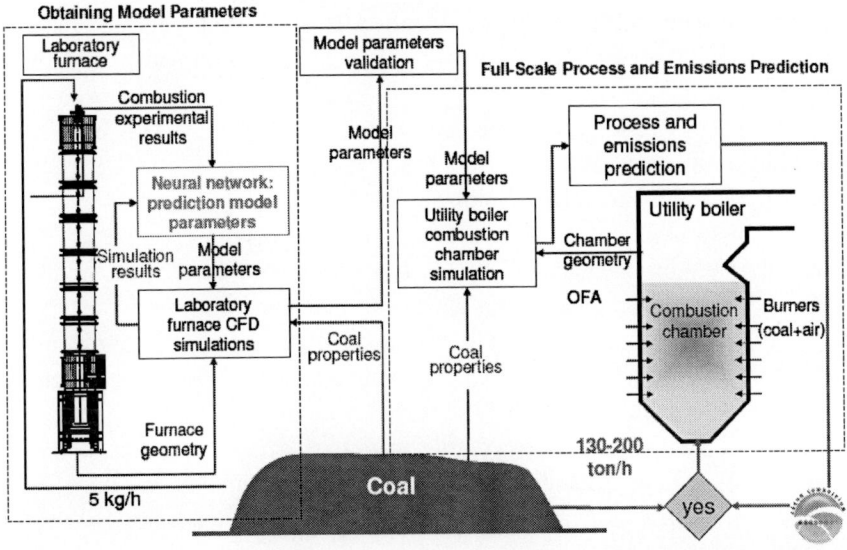

Figure 2. Methodology for predicting coal performance in full-scale utility boilers.

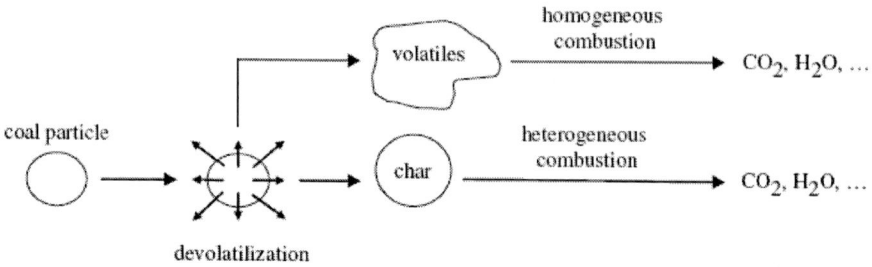

Figure 3. Combustion processes of a coal particle [6].

2.1. KINETIC MODEL PARAMETERS

This chapter describes the combustion kinetics, mechanisms and parameters, used by GLACIER, the commercial CFD program used for this work.

When coal is burned in a combustion chamber, three main processes occur: (1) coal devolatilization, (2) volatiles combustion and (3) char oxidation (char burnout), where figure 3 shows these processes [6]. During devolatilization water evaporates first and then the light gases and tars. These 'volatiles', which are mainly composed of hydrocarbons, oxidize in the combustor environment. The solid material left is rich in carbon, poor in hydrogen and oxygen, but also contains nitrogen, sulfur and mineral matter. This material is termed 'char' and the materials in it continue to react during the process of 'char oxidation' or 'char burnout' [7].

The reaction processes of the coal particles in the GLACIER code include devolatilization, char oxidation and gas-particle interchange [8]. Particles are made up of coal, char, ash and moisture. Ash is inert by definition and volatile mineral matter is considered part of the volatile matter of the coal. Particle swelling is empirically accounted for and due to the small Biot number the particles are assumed to be isothermal. Particle reactions are characterized by multiple parallel reaction rates with fixed activation energies. The off-gas from particle reactions is assumed to be of constant elemental composition. Coal devolatilization is modeled using the two-step model proposed by Ubhayakar et al. [9]. Equilibrium codes for volatile matter oxidation are rather common in combustion modeling and various codes are available in the literature [1,5,10,11]. A global Arrhenius model is used to model heterogeneous char oxidation [8].

Volatile Matter Kinetics

When the coal particle is exposed to the high furnace temperatures, instantaneous devolatilization of the volatile matter (VM) occurs. The volatile matter separates from the char portion of the coal, ignites and oxidizes. The time frame for devolatilization is 1-5 ms and for volatile matter oxidation 50-100 ms.

Devolatilization

Ubhayakar et al. [9] modeled the rapid devolatilization process with two competing first-order reactions, depicted in figure 4. Each reaction illustrates a devolatilization process of the DAF (dry ash free) raw coal, CH_x where $0 \leq x \leq 1$. In the devolatilization process volatile matter, V_1 and V_2, is released from the coal leaving a residual char, R_1 and R_2. The rate constants of these two competing processes are k_1 and k_2. Y_1 and Y_2 are the mass stoichiometry coefficients, which are selected according to the (H/C) ratio of the coal; $Y_n \sim x/x_n$, where $x_n = (H/C)$ of V_n, $n=1,2$. Good agreement between predicted and measured values was achieved using Y_1 as the volatile matter fraction and $Y_2 = 2*Y_1$.

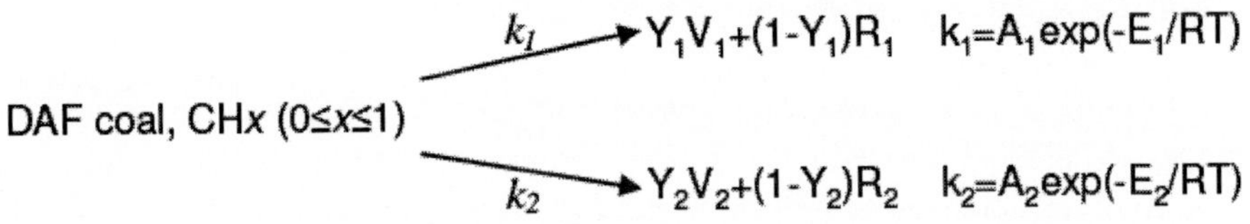

$$\text{DAF coal, CH}_x \ (0 \leq x \leq 1) \xrightarrow{k_1} Y_1V_1 + (1-Y_1)R_1 \quad k_1 = A_1 \exp(-E_1/RT)$$

$$\xrightarrow{k_2} Y_2V_2 + (1-Y_2)R_2 \quad k_2 = A_2 \exp(-E_2/RT)$$

Figure 4. Devolatilization model developed by Ubhayakar et al. [9]: two-step devolatilization model of coal, CH_x ($0 \leq x \leq 1$), DAF stands for dry ash free. k_n are the two rate constants [s^{-1}], Y_n are the mass stoichiometry coefficients, V_n and R_n stand for the volatile matter and residual chars, respectively. For all parameters, $n=1,2$. In the rate constant equations: A_n[s^{-1}] and E_n[kcal/mol] are the preexponential (frequency) factor and activation energy, respectively, R the gas constant and T the temperature of the volatile gases in the vicinity of the coal particle.

The following kinetic parameters for the 2-step devolatilization process are inputted to the GLACIER code: mass stoichiometry coefficients (Y_1 and Y_2), pre-exponential factors (A_1, A_2) and activation energies (E_1, E_2). For a tested blend, the above parameters are entered for each component coal. The instantaneous mass rates of volatile matter production by the competing reactions in figure 4 at time t are assumed as (Eq. 1):

$$\dot{m}_n(r_i,t) = Y_n k_n m_n(r_i,t), n = 1,2 \qquad (1)$$

where $\dot{m}_n$ is the mass of the part of the coal particle with radius r_i that has not yet reacted, the dot represents the time rate of change, $k_n=B_n exp[E_n/RT_c(r_i,t)]$, B_n and E_n are the pre-exponential (frequency) factor and activation energy, respectively, R the gas constant and $T_c(r_i,t)$ the temperature of particles of radius r_i.

The numerical solution to the above model results in a low activation energy reaction significant at lower particle temperatures and a higher activation energy reaction significant at higher temperatures. The high temperature process led to higher yields of lower (H/C) volatiles and char than the low activation process. In other words, a high yield reaction is favored at high temperatures, while a low yield reaction is favored at low temperatures. E_1 and E_2 govern the degree of overall volatile formation as a function of the heating rate. This is the advantage of the two-step devolatilization mechanism, as it predicts the volatile fraction or the changes in the yield as a function of heating rate [12,13].

Volatile Matter Oxidation

Volatile matters oxide very fast in comparison to chemical and physical processes in the furnace; hence it was assumed that once volatile matters are produced, instantaneous equilibrium for gas oxidation reactions takes place. The equilibrium codes are in rather common use in combustion and various codes are available in the literature [1,5,10,11]. One of the most common modeling approaches for volatiles oxidation is the Mixture Fraction approach [5]. Because the reaction time scale is much shorter than the mixing time scale, the physical processes are treated in detail but local instantaneous (infinite rate) chemistry is assumed and computed using an equilibrium algorithm. For two individual inlet streams a conserved scalar variable, f, the mixture fraction, is used to define the level of mixing of primary and secondary mass components, Eq. 2:

$$f = \frac{m_p}{m_p + m_s} \qquad (2)$$

where f is the mass fraction of the primary stream component, m_p is the mass originating from the primary source and m_s is the mass originating from the primary source. Normally, for combustion applications, fuel is taken for the primary stream and the oxidizer is the secondary stream. The advantage of this mixture fraction variable is that any other conserved scalar, s, which is a function of f, such as the fuel density, can be calculated from the local value of f. For instance, Eq. 3:

$$s = fs_p + (1-f)s_s \qquad (3)$$

Therefore the mixture fraction approach lowers the number of conserved scalars required to describe a combustion system. The validity of this approach requires that the turbulent diffusivity and boundary conditions of all gas phases be the same, which is not unreasonable in most large combustion systems. Statistical probability density functions (pdf's) are used at each point in the flow field to obtain the mean properties of chemical species, temperature and flow dynamics based on the local instantaneous equilibrium calculated with the mixture fraction approach. The oxidation kinetics, with fixed activation energies, are pre-programmed into the GLACIER CFD code and cannot be changed by the user.

Char Kinetics

The char oxidation rate combines the effects of surface reactivity and pore diffusion as described by Hurt and Mitchell [14] and shown Eq. 4:

$$q = k_s P_s^n \qquad (4)$$

where q is the combustion rate normalized by the particle external surface area, k_s is the global rate coefficient [kg-Carbon/m^2-s], given by Eq. 5:

$$k_s = A \exp(-E/RT_p) \qquad (5)$$

A [kg-Carbon/m^2-s] represents the global pre-exponential factor, E [kJ/mol] the global activation energy, R the gas constant and T_p [K] the temperature of the char particle. P_s is the partial pressure of oxygen at the particle surface and n is the global reaction order. Three of the above parameters governing char oxidation can be varied in GLACIER: n, A and E.

NO_x Kinetics

Most coals contain 0.5-2.0% nitrogen by weight. Bituminous coals generally have high nitrogen levels and anthracite low nitrogen levels. The large fuel-nitrogen content of coal can result in substantial NO_x emissions. The formation and destruction of NO_x in combustion systems is very complicated. During combustion, nitrogen found in the coal or in the combustion air is transformed by many different chemical mechanisms to nitrogen-containing species such as: nitric oxide (NO), nitrogen dioxide (NO_2), nitrous oxide (N_2O), ammonia (NH_3), hydrogen cyanide (HCN) and amine compounds. Coal combustion systems emit nitrogen oxides mostly in the form of nitric oxide (NO), with a small part appearing as nitrogen dioxide (NO_2) [15,16].

Fuel-nitrogen is distributed between the volatiles and the char and it is released during devolatilization and char oxidation. This split is potentially important for NO_x formation. The fraction of nitrogen released with volatiles during devolatilization depends on the fuel type, the temperature, the heating rate and the residence time [17]. NO_x formation increases with the oxygen content in the coal, meaning that a high rank coal will form less NO_x. Increasing the temperature and the residence time favors the conversion of coal-N to volatile-N. In regular pulverized coal combustion, the source of 60-80% of the NO_x is volatile-N. Because of its early release, the volatile matter of the coal has a major impact on the amount of NO formation. The heating rate and final temperature of the coal particle most probably influence the devolatilization processes, and, as a result, the amount of nitrogen released from the coal. Temperatures above 2000 K are required for complete char-N release. The magnitude of char-NO_x formation is dependent on char properties, reactivity and internal surface area, and the combustor environment; temperature and specie fields [18]. Several factors determine these characteristics of the char: fuel-type, air-staging, burner injection and the thermal and physical conditions in the combustor. Basically, in laboratory conditions, one will find that the contribution of NO_x from char combustion is less than that from volatiles combustion. However, the homogeneous NO_x reactions are influenced more by combustion modifications than the heterogeneous NO_x reactions. Therefore, in "real-life" facilities, the control of NO_x formed from heterogeneous reactions is more difficult. Spinti and Pershing [19] concluded that char-N conversion to NO_x, for typical pulverized coal combustor conditions, is 50-60% with low rank coals and 40-50% with bituminous coals. According to Williams et al. [20] char-N amounts to about 80% of the total NO formed. Rostam-Abadi et al. [21] give a lower amount of 20-30% and state that the amount is dependent on the temperature and the amount of devolatilization.

The type of nitrogen species released during the devolatilization process is determined by the structure of nitrogen in the coal. Coal nitrogen is believed to occur mainly in aromatic ring structures. At high temperatures (>1500 K), 70-90% of the coal nitrogen is devolatilized with 0-20% of the coal nitrogen evolving in the early volatiles, mainly as HCN and NH_3. The tars and nitrogen bound in aromatic rings are the main source of HCN formation while the amines in the coal mainly form NH_3. Combustion of bituminous coals show more HCN formation than NH_3 and sub-bituminous and lignite coals form more NH_3. This is probably because the number of aromatic rings decreases with coal rank but the number of cyclic compounds increases. Accordingly, in low-rank coals the conversion of fuel nitrogen to NH_3 increases [16]. In another study by Schafer and Bonn [22] it was established that the concentration of NH_3 is usually higher than that of HCN in the combustion chamber. These larger amounts of NH_3 cannot be explained only by conversion of the coal-amines but most probably are formed from hydrolysis of HCN. Knowledge of the HCN and NH_3 concentrations is very important since these are the main precursors of nitrogen oxides.

It must be stressed again that the temperature, residence time and fuel/oxygen ratio are main factors in determining the species formed. The three main mechanisms for NO_x formation in combustion systems are [23]: (1) Thermal NO_x - the reaction mechanism that occurs at high temperatures between atmospheric nitrogen and oxygen, (2) Prompt NO_x - the attack of a hydrocarbon free radical on molecular nitrogen producing NO precursors, and (3) Fuel NO_x - the oxidation of the fuel-nitrogen released during devolatilization and char oxidation. Fuel NO is the main source of NO_x in coal-fired systems and accounts for 60-80% of the total NO_x formed [24].

Fuel-N is assumed to proceed through HCN and/or NH_3 [16]. When the fuel-N is bound in an aromatic ring, HCN is formed. NH_3 is formed from amines. Figure 5 presents the reaction paths for fuel NO_x. Nitrogen in the volatile matter is released during the devolatilization process. Part of the nitrogen (α) is quickly transformed to HCN and the remaining part ($1-\alpha$) of the fuel nitrogen is transformed to NH_3. Depending on local conditions, these two species will react to form either NO or N_2. They will be reduced to N_2 in fuel-rich regions, and in fuel-lean regions they are generally oxidized to NO_x. The

NO_x formed can also be reduced via heterogeneous reactions with the char particles.

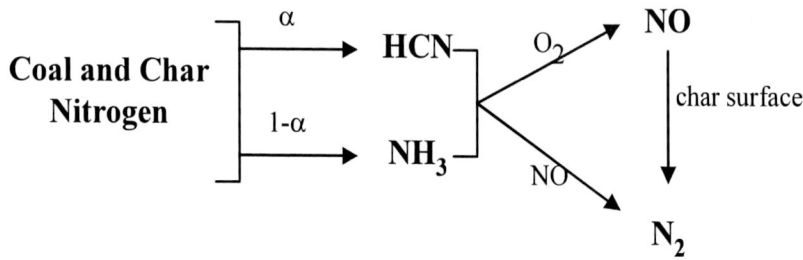

Figure 5. Schematic representation of fuel NOx formation and reduction process [modified from Hill and Smoot [16]].

The GLACIER code used in this work, as most coal combustion CFD models, separates the mechanisms of NO_x formation from the generalized combustion model. The models of nitrogen pollutants are executed after the flame structure has been predicted. The basis for this is that the flame structure is governed by much faster fuel-oxidizer reactions and is not affected by the formation of trace pollutant species. An advantage of this approach is computational efficiency. The time required to solve the system of equations for the fuel combustion can require many hours of computer time while the pollutant sub-models typically make up a small part (~10%) of the time required to converge the combustion case. Thus, NO_x sub-model parameters and pollutant formation mechanisms can be better investigated by solving the NO_x sub-model using a pre-calculated flame structure [16].

GLACIER cannot compute the full chemistry for all intermediate, including nitrogen, species. However, this detail is not required for engineering solutions for utility boiler combustion chambers. As long as all physical mechanisms of first-order importance are included in the investigation, engineering modifications to the furnace can be simulated using a condensed set of chemical kinetic mechanisms [8]. For NO_x formation these are governed by three variables that are used for NO_x postprocessor simulation only. These parameters, described therein, are varied by the user at the beginning of each NO_x simulation and are inputted into the CFD code. The variables used for the NO_x postprocessor are named VMNFR, ZEDA and ZEDAH. The following description was summarized from the GLACIER manual [Reaction Engineering International (REI), "Configured Fireside Simulator IEC 550 MW Wall-Fired Unit" (2004)].

VMNFR is the parameter that defines the division of nitrogen between the volatile matter and the char, and it is experimentally derived as described in Chapter 2.2. This split is potentially important for NO_x formation. The default value for VMNFR is 0.5; equal mass fractions of nitrogen in the volatile matter and the char. The absolute amount of nitrogen can be different in volatile matter and char and is dependent on the volatile yield. For VMNFR = 1.0, all coal-N is in the volatile matter. For VMNFR = 0, all coal-N is in the char. If X symbolizes the volatile matter yield expressed as a fraction of total coal organic mass, for $0.0 \leq VMNFR < 0.5$ there is less coal nitrogen partitioned into volatile matter nitrogen (X*(VMNFR/0.5)) than into char nitrogen (1-(X*(VMNFR/0.5))) relative to equal mass fractions of nitrogen in the volatile matter and char (VMNFR=0.5). For $0.5 < VMNFR \leq 1.0$ there is more coal nitrogen partitioned into volatile matter nitrogen (X+((VMNFR-0.5)/0.5)*(1-X)) than into char nitrogen (1-(X+((VMNFR-0.5/0.5)*(1-X))) relative to equal mass fraction of nitrogen in the volatile matter and char (VMNFR=0.5).

ZEDA is the parameter that partitions the formation of volatile-N between HCN and NH_3. Knowledge of the HCN and NH_3 concentrations is very important since they are the main precursors of nitrogen oxides. ZEDA is defined by:

$$ZEDA = \frac{mol_{HCN}}{mol_{HCN} + mol_{NH_3}} \qquad (6)$$

The default value for ZEDA is 0.8; meaning 80% of the volatile-N forms HCN. ZEDAH is the parameter that defines the fraction of char-N that is converted to NO_x. The CFD default value is 0.1; meaning 10% of the char-N is converted to NO_x. It should be noted that this value is low compared to the literature [19-21].

Since the formation of NO_x is so closely coupled with operation conditions: fuel/air stoichiometry, temperature, heating rate and residence time, the NO_x postprocessor parameters are likely to change with different operating conditions. This differs from the volatiles and char oxidation kinetic parameters which are kept constant for each coal.

Summary

Errors in the model parameters affect combustion characteristics such as coal burnout, temperature fields, flame structure and NO_x formation. Lockwood et al. [25] carried out a comprehensive sensitivity study on their test furnace on the influence of different model parameters on combustion performance. They show that although the volatile yield did not have much of an effect, devolatilization kinetics influence is strongly felt in the near burner zone and volatiles combustion rates strongly influence oxygen levels for their furnace length. Char oxidation kinetics govern the burnout rate. Gera et al. [26] showed that deviating the 1-step devolatilization activation energy by ±12.5% moves the flame root position (defined as the distance from burner where flame temperature reaches 1000 K) in the range of l/D (length/diameter)= 0.69 to 1.04. Jones et al. [13] demonstrate that the devolatilization rate has a marked influence on predicted NO_x emissions. Sheng et al. [10] found that the total coal nitrogen content and volatile yields influence the levels of NO_x formation. Kurose at al. [27] demonstrated that taking into account NO formation from char-N improved their prediction results.

To summarize, kinetic data required for running GLACIER are: nitrogen distribution between volatile matter and coal-char (VMNFR), six parameters for the two-step devolatilization mechanism ($Y_{1,2}$, $A_{1,2}$, $E_{1,2}$), three parameters for the one-step coal oxidation process (n, A_c, E_c), a set of kinetic parameters (plugged-in the code) for a comprehensive gas-phase NO_x mechanism starting from HCN-NH_3 with an initial ratio (ZEDA), one parameter for the conversion of char-nitrogen into NO (ZEDAH); a total of twelve parameters, that none of them could be found in the literature. All of these parameters which describe the coal particles' combustion and NO_x formation rates can be varied in the GLACIER program to obtain the best agreement with the experimental data.

The approach described here determines a simplified kinetics model. While it cannot be said that the kinetic parameters used in the model will correspond to the combustion rate of a single particle of coal, these parameters do describe the combustion behavior of a "macroscopic" sample of the tested coal. For the goal of predicting the combustion behavior of these coals and blends in a utility boiler furnace, this simplified model used with the GLACIER code gave good results.

From the previous chapter one understands the complexity of modeling coal combustion. The CFD code used in this study, GLACIER, allows the user to define twelve different parameters that influence the different combustion species, combustion rates and temperatures. These parameters are varied in the

GLACIER code to obtain the best agreement with the test furnace experimental data for further input in the GLACIER code of full-scale utility boilers.

2.2. TEST FACILITY

The core of the facility is the 50 kW cylindrical down-fired test furnace. As the foremost function of the furnace is to extract kinetic parameters for the large-scale boiler simulations, plug-flow conditions in the furnace are essential, i.e. velocity, concentration and temperature are uniform throughout the radial coordinate. Plug-flow conditions for most of the furnace length were achieved by two means using: (1) a high-swirling burner at the top of furnace to ensure immediate and full mixing of fuel and air and (2) a relatively high length(L)/diameter(D) ratio (L=4.5 m and D=0.2 m; L/D=22.5). To recreate utility boiler kinetic conditions as much as possible, the test furnace works at temperatures up to 1650 °C, corresponding to the temperatures at pulverized-coal facilities. It should be emphasized that the test facility does not simulate the flow dynamics in the utility boiler and that the test facility cannot directly predict the combustion behavior in utility boilers. Coal combustion processes in test facilities, such as this one, are simplified because many of the physical and chemical processes that take place in a utility boiler are absent. Therefore, it is easier to validate CFD combustion models for the test facility and apply the validated model, with the same kinetic parameters but with different boundary conditions, to a utility boiler.

Process Design

Figure 6 describes the test facility's process scheme. The system is designed to burn solid, liquid and gaseous fuels and can burn one or two types of pulverized coals and different ratios of binary blends. Different stoichiometric ratios of air-fuel can be determined; air and fuel staging are done. The hot combustion gases are cooled before being emitted into the atmosphere and the particulate matter is separated using a bag filter. The whole process is semi-automatic and controlled by a fully computerized control system using about fifty strategically placed sensors.

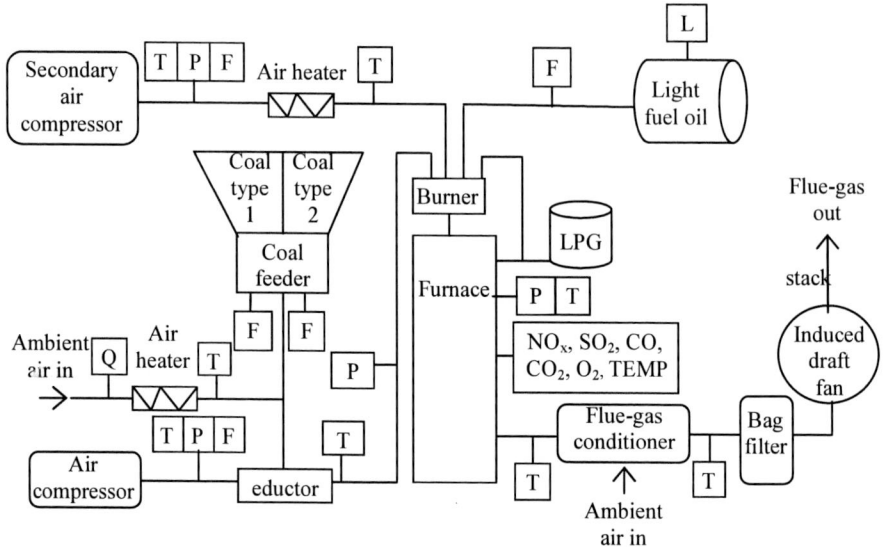

Figure 6. Test facility process scheme. T, P, F, W, L denote measurements (manual and automatic) of temperature (°C), pressure (bar), air and fuel flow (air, F=Nm³/h), fuel oil (F=L/h), LPG (F=L/min), coal (W=kg/h)) and level of fuel oil in tank respectively.

Coal Preparation and Feeder

Pulverized coal is prepared on site from raw coal with a size distribution of about 40% smaller than a 5 mm diameter, depending on the coal. The coal is air dried for 24 hours, crushed with Mini Crusher 8F and then pulverized with Laboratory Pulverizer LC67, both from Gilson Inc., USA. Normally the pulverized coal particle size distribution is 60-80% smaller than 75 micron, but this can be changed as needed for the combustion experiments. The coal dosing structure, shown in figure 7, supplies the pulverized coal to the primary air line. It was planned and built specifically to meet the requirements of the test facility. The coal dosing system has the capability of supplying a homogeneous mixture of two types of pulverized coal at a rate of 2-8 kg/h. The coal is gravitationally transferred directly from the drum to the loading hoppers which can each hold a different type of coal. A vibrator periodically shakes the loading hopper to eliminate the formation of bridges or clumps. When the controller relays an order, coal is dropped from the loading hopper to the weighing hopper. An agitator is installed in the weighing hopper to mix the coal and break up lumps.

After the coals are weighed, they are transferred to a single outlet and are released to the primary air-line via an ejector.

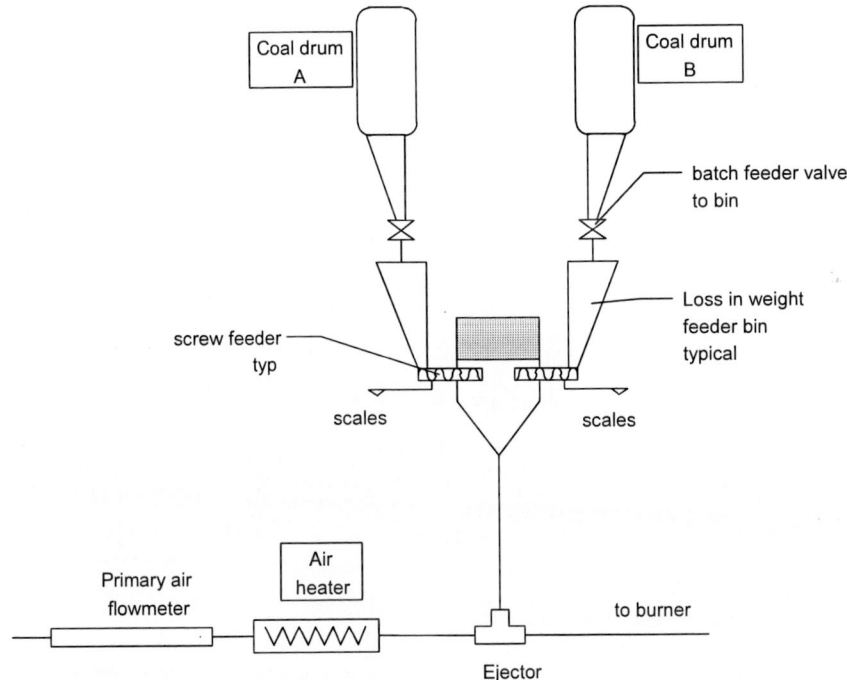

Figure 7. Coal feeder system.

Burner

A highly-swirling burner was designed to create an intense and short mixing region. As shown below, the burner is located at the top of the furnace. Figure 8a shows the cross section of the burner and figure 8b shows the burner face.

The burner is made from stainless steel and is insulated from the heat of the furnace by a quarl shaped protective shield. Fuel oil or liquid petroleum gas (lpg) can be injected through an appropriate nozzle located in the center of the burner. Surrounding the nozzle is an open ring (d=0.020/0.022 m) for air flow, whose purpose is to reduce recirculation in the central zone of the burner exit when burning fuel oil. Without this air flow, dendrites of soot form on the burner center. Surrounding this ring are nine holes through which the pulverized coal and primary air mixture enter the furnace. The total area of the coal-air inlet is

2.54E-04 m² while each hole has an ID of 0.006 m. The holes are located on an imaginary circumference of 0.035 m. Secondary air is introduced via twenty-four equal rectangular holes, with the dimensions of 0.0055x0.0035 m, total area 3.76E-04 m². These holes are arranged circularly with an OD (outer diameter) of 0.057 m and ID (inner diameter) of 0.047 m and they are machined at 45° with the furnace centerline, to cause a swirling motion.

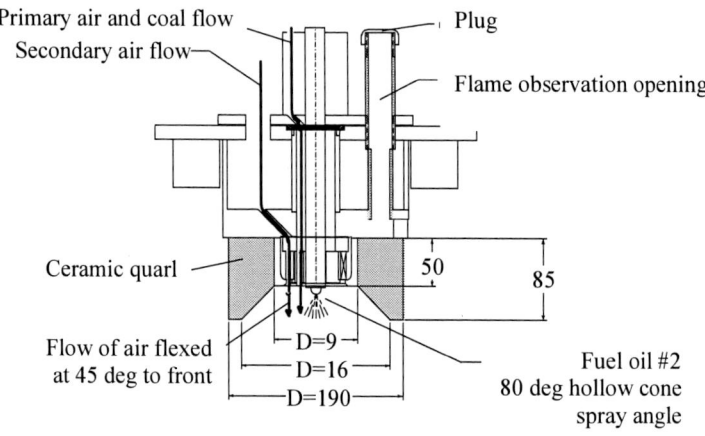

Figure 8a. Cross section of the burner.

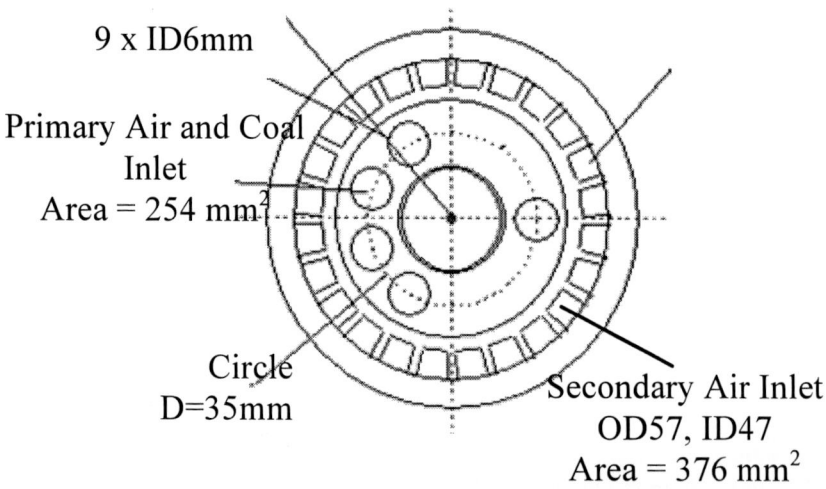

Figure 8b. Burner face.

Figure 9 shows the axial component of the velocity vector (the velocity in the direction of the furnace length) obtained by the CFD simulations done with the GLACIER code. Because of the grid layout, the vectors are denser in the center of the furnace. As expected from the burner design, the flow is complex with recirculation zones near the burner, but by 0.3 m from the burner face, at l/D (length/diameter) = 1.5, the axial flow pattern is quite uniform for the length of the furnace. In the near-burner zone one can see the recirculation of the flow along the sides of the furnace near the quarl and in the center, back towards the burner face.

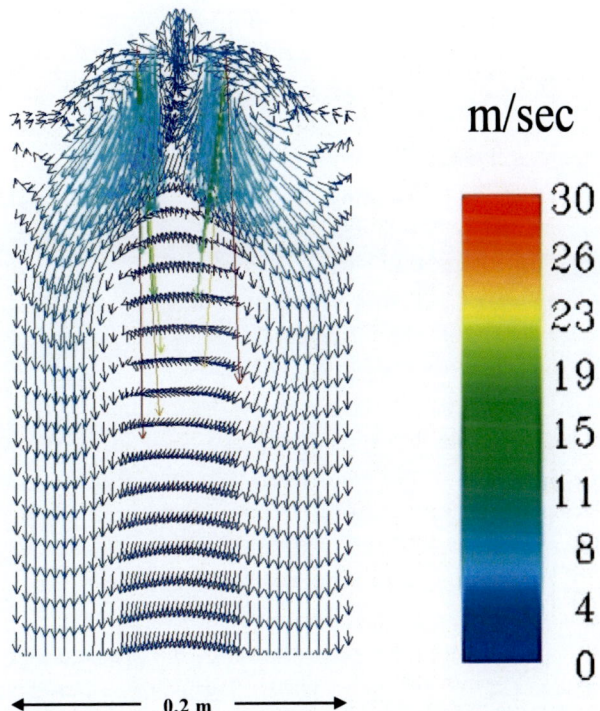

Figure 9. Axial vector presentation of the velocity in the near-burner zone (l/d = 0-2).

Test furnace

Figure 10 shows the test furnace, which dimensions are 0.2 m id (inner diameter) and 4.5 m high. The combination of a high-swirling burner, small furnace diameter and long furnace length creates a short mixing region followed by a plug-flow in the rest of the furnace. This will be explained in greater detail after the furnace description. The furnace can work up to temperatures of 1650 °C, and has a residence time of 1-2 sec. These correspond to the temperatures at pulverized-coal facilities and the time needed for complete coal combustion. The furnace is made up of 8 modular sections, each around 0.6 m high. The cast ceramic center is surrounded by four insulation boards to minimize heat losses. The outer shell temperature decreases monotonically from 85 °C (top furnace) to 45 °C (bottom), depending on the furnace load. For safety and stable flow considerations the furnace is maintained at a slightly sub-atmospheric pressure (about 0.5 mbar below 1 atm). The primary air is heated to about 110 °C with a flow rate around 12 Nm^3/hr, which is the minimum rate required to prevent the coal from settling in the pipe. The coal and primary air mixture enters the burner at a temperature of 60-70 °C. The secondary air is heated to 250 °C, with flow rate depending on the feed rate of the fuel and stoichiometric ratio. Twenty, 0.04 m id, sampling openings were drilled on both sides of the walls of the furnace for gas and temperature sampling, and other input and output applications. Overfire air (OFA) is added via 0.008 m id openings at l/D=13.83. LPG injection is through 0.0006 m id openings at l/D=4.83, 7.83 or 10.83.

The specific geometry (2-D with small ID) of the furnace serves two purposes: (1) simplify the CFD simulations, (2) create a short mixing region followed by a plug-flow for most of the furnace length. The high-swirling burner causes a short and yet very intense mixing region. In this mixing region, under certain conditions, both volatile matter and char are burned. For the rest of the furnace chemical reactions are minimal unless reburning fuel or OFA were added. This two-zone configuration was obtained both from simulation and experiment, as shown in figures 11-15. The lengths of these two zones were varied by changing different combustion process parameters, such as feed particle size and stoichiometric ratio.

Figure 11 shows two-dimensional presentation of the predicted temperature within the test furnace (125 °C between each isotherm-line), distributed in five furnace segments (0.6m for each segment, or 3 l/D). The temperature is not uniform near the burner, the intense mixing zone, and subsequently, by l/D=4, or about 1 m from the burner face, the temperature is homogeneous for the rest of the furnace length. Predicted oxygen levels, depicted in figure 12, show similar trends. Figure 13 portrays the axial velocity component (the velocity in the direction of the flow) distribution in the furnace. The flow pattern is varied near the burner but by the middle of the first segment, l/D=1.5, the flow pattern is uniform and stays that way for the length of the furnace. The radial profile of the velocity shows higher speed in the furnace center, and almost no movement near the furnace walls. Since the burner causes high-swirling, there is radial velocity near the furnace walls, which is not depicted by this figure.

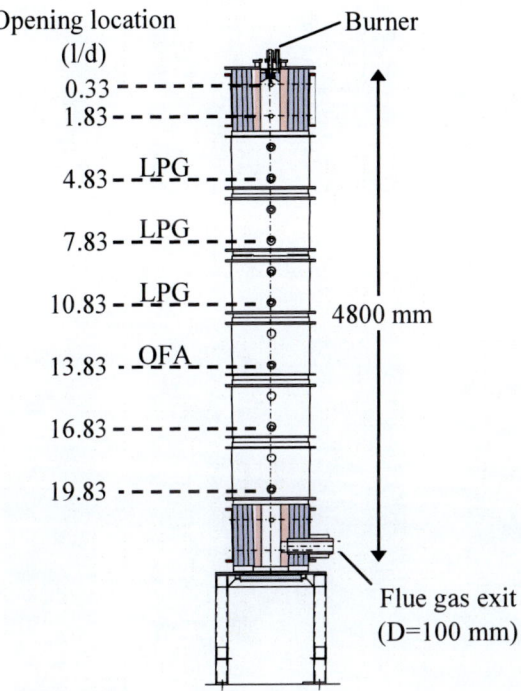

Figure 10. Furnace – general view and sampling, LPG and OFA openings locations.

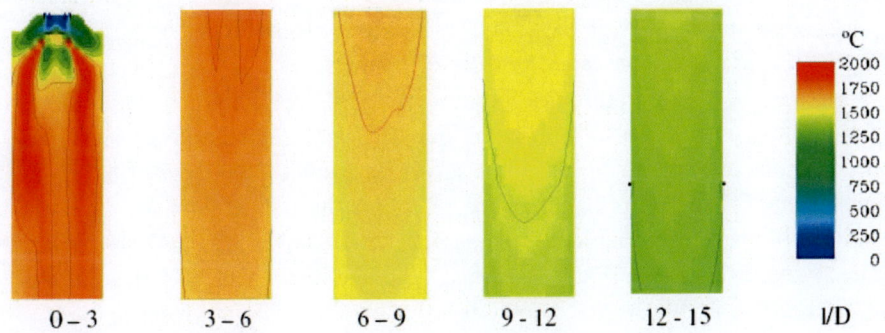

Figure 11. Two-dimensional presentation of the temperature within the test furnace, with isotherms (125 °C between each line), distributed in 5 segments for clarity, each segment represents 3 l/D.

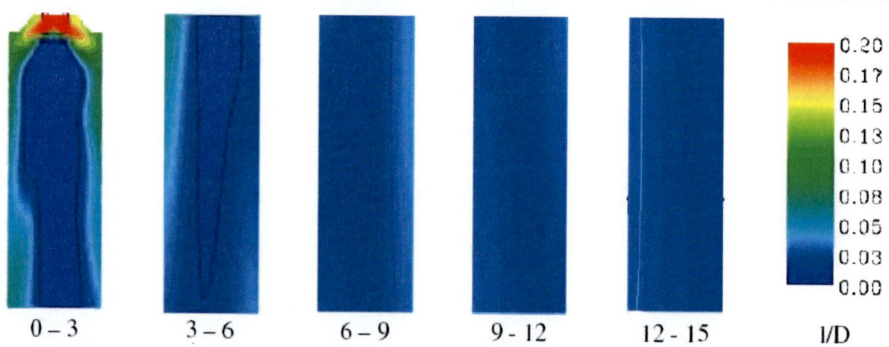

Figure 12. Two-dimensional presentation of oxygen levels within the test furnace distributed in 5 segments for clarity, each segment represents 3 l/D.

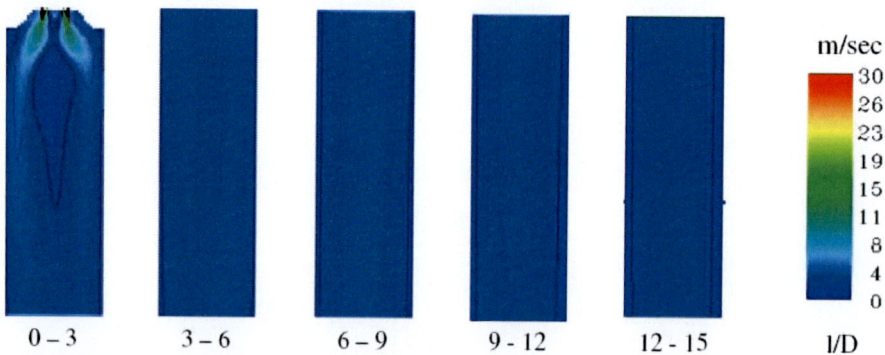

Figure 13. Two-dimensional presentation of the axial velocity within the test furnace distributed in 5 segments for clarity, each segment represents 3 l/D.

The two-zones, high-mixing and plug flow, were also experimentally observed. Figure 14 shows radial profiles of temperature and gas concentrations at l/D=0.33 and 1.83 for three different coals: KPC (bituminous), Drummond (bituminous) and Adaro (sub-bituminous). The temperatures and gas concentrations are not uniform at these locations, which is the area of the high-mixing zone. On the other hand, figure 15 shows the temperature and gases in the plug-flow region by depicting temperature and CO_2 at l/D=4.83, 10.83 and 13.83. These are quite uniform. It should be noted again that both simulations and measurements show the two-zone configuration of the furnace that is essential for the extraction of kinetic parameters.

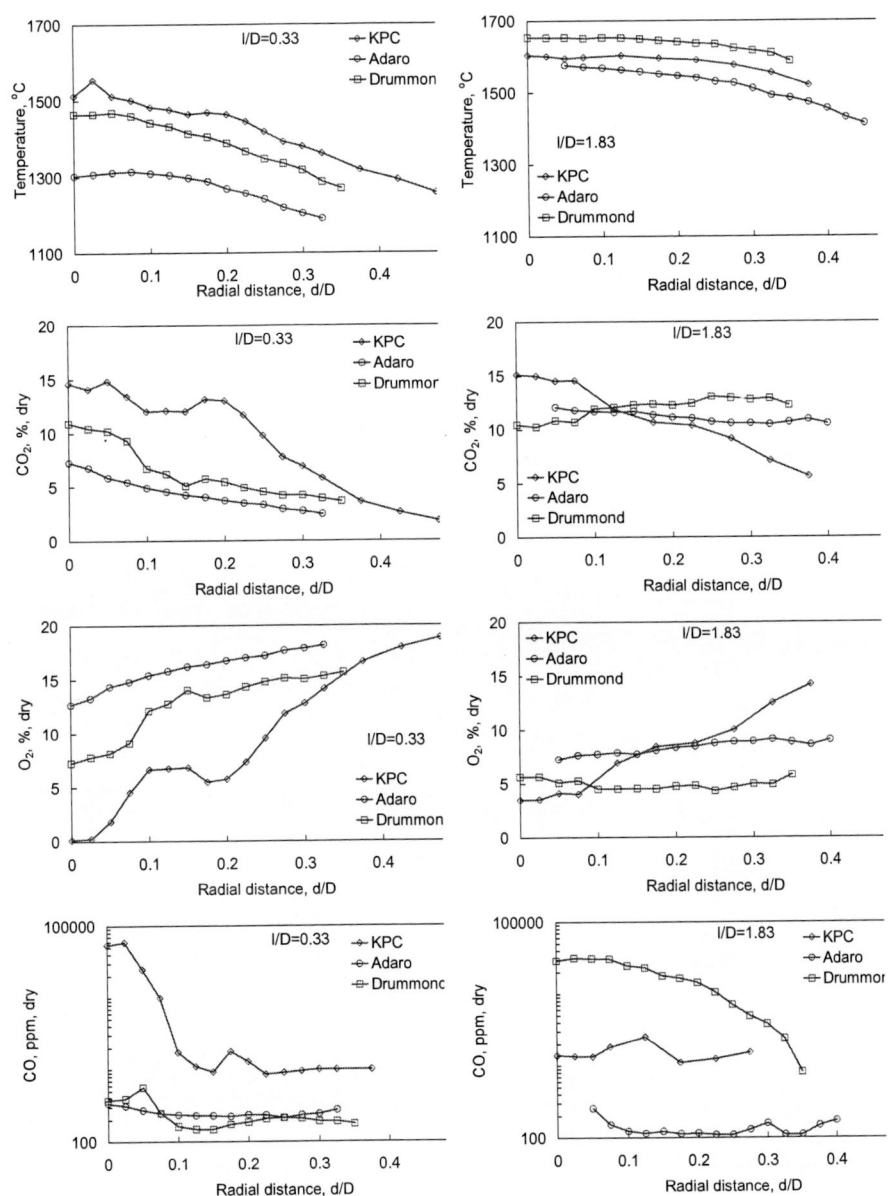

Figure 14 continued on next page.

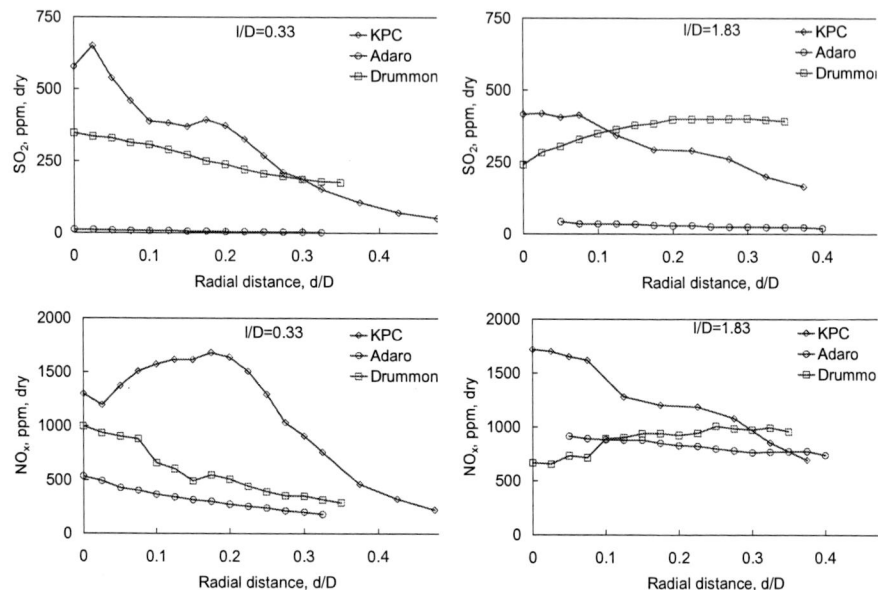

Figure 14. Measured radial profiles of temperature and gas concentrations in the high-mixing zone at l/D=0.33 and 1.83 for KPC, Adaro and Drummond coals.

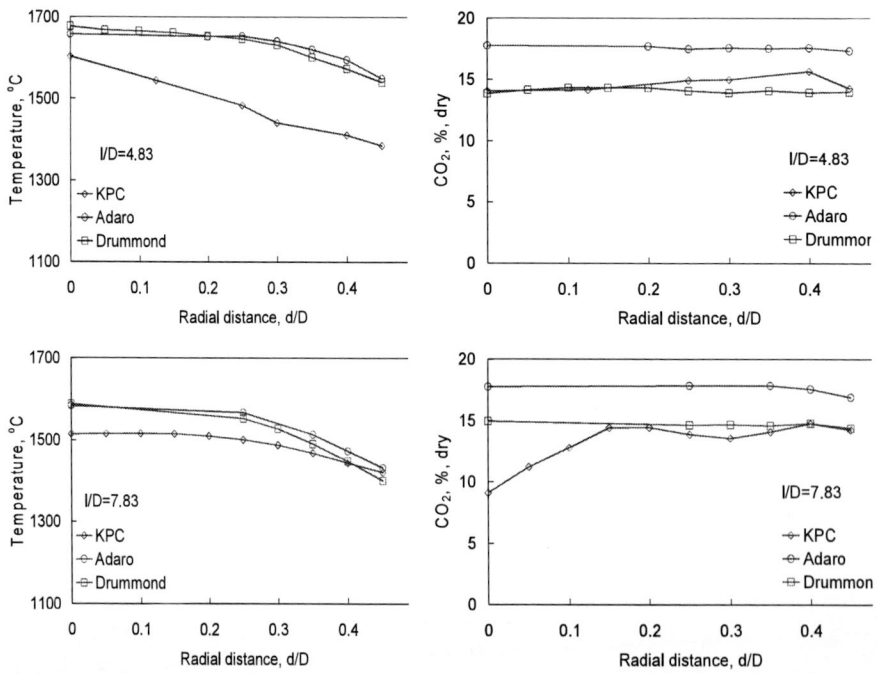

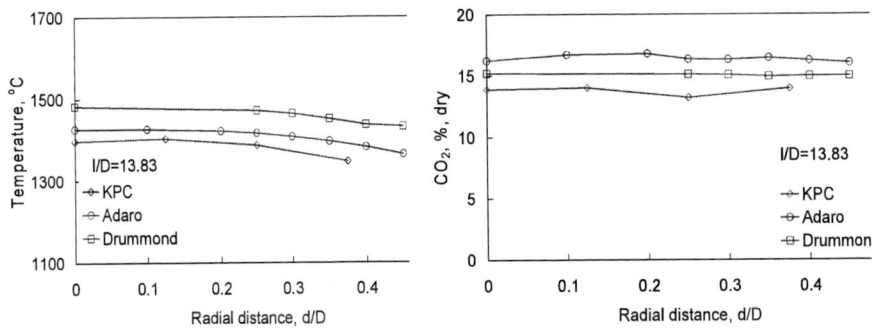

Figure 15. Measured radial profiles of temperature and CO_2 concentrations in the plug-flow zone at l/D=4.83, 7.83 and 13.83 for KPC, Adaro and Drummond coals. CO_2 represents the behavior of all other gases measured in the furnace.

Sampling and Analysis

The temperature and gas concentration monitoring in the test furnace is done semi-automatically. The probes are moved manually but the data is collected and analyzed electronically and stored by the control and data acquisition system. Care was taken to ensure that both temperature and gas concentrations results were reproducible by periodically checking the stability of the test facility and the measuring equipment.

Temperature

Temperature measurement in combustion systems is a very complex issue, especially for coal combustion, where particles are present. Particle and gas temperatures are measured separately. Particle temperatures were not measured in this work. Gas temperature was monitored with thermocouple probes. These include bare fine wire thermocouples and aspirated thermocouples, like the suction pyrometer. The main advantages of bare fine wire thermocouples are that the measurements have the high precision characteristic of electrical measurements; the dimensions are small so that flame disturbance is minimal, and the method is simple and inexpensive.

However, there are several sources of error when using thermocouples in coal combustion, mostly radiation effects. The thermocouple does not filter out the radiative heat transfer from the combustor walls and conduction heat transfer along the leads. Radiative heat transfer between the particles and the bead, and coating of the bead by particles, are sources of error when particles are present. Therefore, the thermocouple will not obtain the same temperature as the hot gas. In suction pyrometers, refractory shields minimize the effects of radiant heat by shielding the thermocouple from its surroundings. The suction pyrometer's drawbacks are its' bulky size and its' requirement of high suction speed (200 m s^{-1}), which can disturb the flow, but mainly that it clogs up very easily because of its small diameter. To solve this problem, the probe was cleaned every 10-15 min or, as needed, with a pulse of pressurized air. This prevented obstruction of the probe but caused cracking of the ceramic (Alsint >99% silica) shield. It was decided to calibrate the thermocouple connected to the gas probe with the suction pyrometer and to use this temperature measurement on a regular basis. As expected, a strong linear correlation between the two probes was found (see figure 16). A literature correlation from Spinti [28] was also used for this calibration. TypeB (0-1700 °C), 0.50 mm diameter, thermocouples were used.

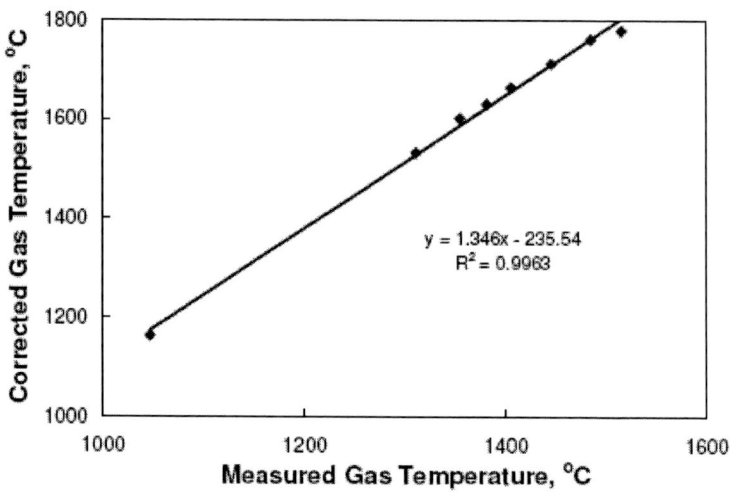

Figure 16. Correlation between corrected temperature (using suction pyrometer and literature correlation) vs. measured gas temperature with gas analyzer probe.

Combustion Gases

The main components of the system for sampling and analyzing gases are (1) gas sampling probes, (2) heated gas sampling tube, particulate filter unit and water condenser, (3) NO_2 to NO converter, and (5) Gas analyzers (CO, CO_2, NO_x, SO_2 and O_2). The ideal gas probe provides a sample that is representative of the gas species' concentrations at the point in the combustor where the probe tip is located. The designs of most sampling probes are based on convective cooling, water or gas quenching or aerodynamic quenching techniques. The quenching of gas samples extracted from combustion systems is essential for the prevention of chemical reactions occurring in the probe or intermediate lines that may significantly change species concentrations. Figure 17 shows the design of the water-cooled probe constructed for this project.

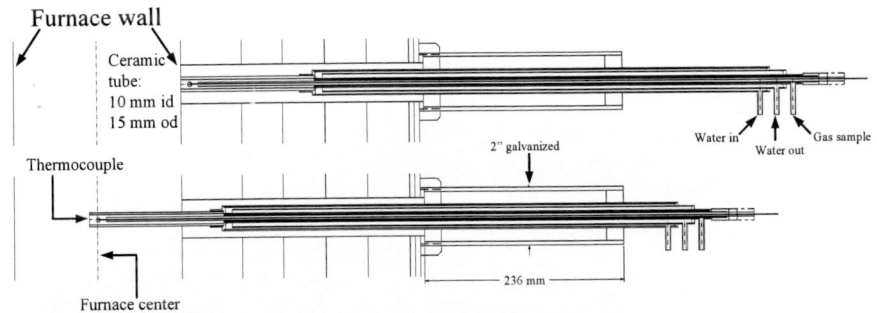

Figure 17. Probe (water-cooled) for temperature measurement and combustion gas extraction.

The sampling tube is Teflon made 7-meter long, 5mm ID, with flow range of 30-300 L h^{1}. The tube is electrically heated to 130 °C to prevent water condensation in the pipeline; this is to prevent gas absorption and chemical reactions between the gases and water, which will subsequently lower the gas concentrations. Once the gas reaches the analytical unit, it is cooled and the condensate is separated and pumped out of the unit. A ceramic filter, installed between the gas probe and the analyzer system cabinet, traps the particulates. The capacity of the filter is 0.010 kg m^{-3} and its' retention rate is 99.99% for particles >5μm. The filter is checked and cleaned periodically with compressed air. This cleaning is important, not only to prevent clogging, but also to lower, as much as possible, the absorption of gas on the particulate matter.

The combustion gases analyzers system is installed in a closed cabinet and works in the temperature range (−20)-50 °C. Analyzer calibration is done on a daily basis with built-in permeation tubes. Once a year, calibration is done by an external laboratory.

Table 1. Measurement parameters of CO, CO₂, NO and SO₂ (two analyzers)

Component Parameter	CO	CO_2	NO	SO_2
Range	0-1%, 0-10%	0-30%	0-2,500 ppm	0-10,000 ppm
Linearity deviation	≤ 1% of span	≤ 1% of span	≤ 1% of span	≤ 1% of span
Repeatability	≤ 0.5% of span	≤ 0.5% of span	≤ 0.5% of span	≤ 0.5% of span
Zero drift	≤ 3% of span/week	≤ 3% of span/week	≤ 3% of span/week	≤ 3% of span/week
Sensitivity drift	≤ 1% of measured value per week	≤ 1% of measured value per week	≤ 1% of measured value per week	≤ 1% of measured value per week
Output signal variations	5 sec	5 sec	5 sec	5 sec
Detection limit	≤ 0.5% of span	≤ 0.5% of span	≤ 0.5% of span	≤ 0.5% of span

Table 2. Measurement parameters of O₂ (one analyzer)

Range	0-25 Vol.-% O_2
Linearity deviation	≤ 0.5% of span
Repeatability	≤ 0.5% of span
Zero drift	≤ 0.03% of span per week
Sensitivity drift	≤ 1% of measured value per week
Output signal variations	3 sec
Detection limit	≤ 0.5% of span

Monitoring is done with three gas analyzers that monitor the following components in the furnace: (a) CO, CO_2, NO and SO_2 with Infrared Analyzer Module Uras 14 (two analyzers). (b) O_2 Analyzer Module Magnos 106. The Uras 14 uses the fact that different types of heteroatomic molecules absorb infrared radiation at specific wavelengths and utilizes this principle of non-dispersive infrared radiation (2.5-8 μm wavelength range). Each instrument measures two components (CO/CO_2 and NO/SO_2). Table 1 summarizes the measurement parameters for each gas. Since the analyzer cannot recognize NO_2, in order to get an analysis of NO_x (NO+NO_2) the gas sample needs to be diverted to a converter, which reduces the NO_2 in the gas sample to NO. The conversion is performed in a reaction tube filled with a catalyst (molybdenum) and heated by the tubular furnace. If the gas sample is not put through the converter then the analyzer will measure only the NO concentration in the sample. The Magnos 106 measures oxygen concentration by utilizing oxygen's paramagnetic behavior. Table 2 summarizes the oxygen measurement parameters.

Heat Flux

Heat flux is created by a temperature gradient and is the rate of energy transfer through a given surface. This quantity can be measured using a heat flux sensor. Generally, a heat flux sensor is made up from a plate with a differential temperature sensor connected to the top and bottom, at different temperatures. The thermocouple generates an output voltage that is proportional to the temperature difference between the hot- and cold sides. Assuming that the thermal conductivity of the plate is constant and the flow is static, the heat flux is proportional to the measured temperature difference. Figure 18 shows the heat flux probe. The potential difference between the hot and cold leads of the thermocouple (TC) is measured. This potential is translated to heat flux by the equation given in figure 19. The probe was calibrated with a calibrated IEC probe to obtain the graph shown in figure 19. The central thermocouple temperature measurement is important for proper heat flux measurement, because the heat flux is temperature dependent.

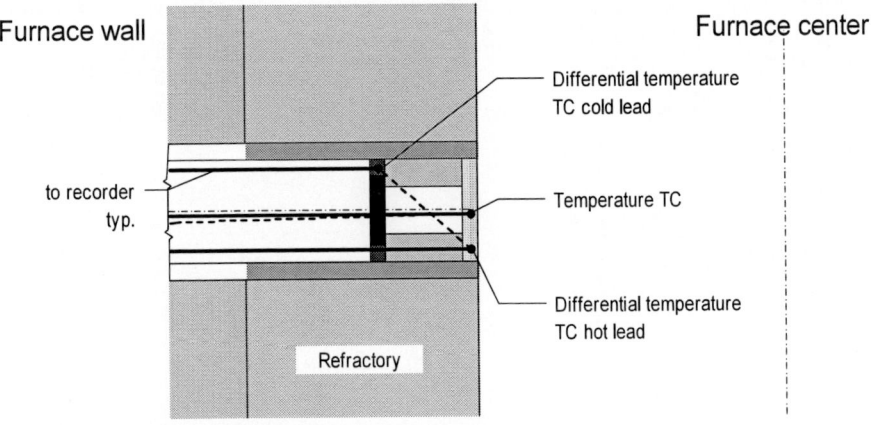

Figure 18. Heat flux probe layout.

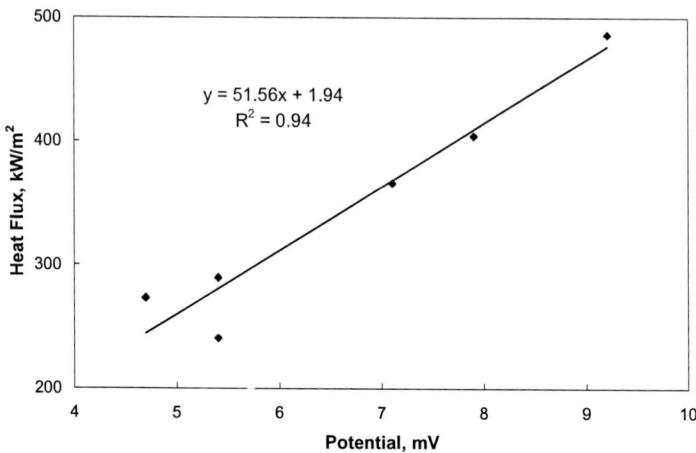

Figure 19. Heat flux probe calibration and equation for potential/heat flux transformation.

LOI (Loss-on-Ignition)

Loss-on-ignition, LOI, is the amount of carbon in the combustion particulate residue (mostly ash). Obviously, low levels of LOI are the goal of the utility boiler operator. LOI levels of 2-5% in the combustion chamber are considered normal and vary according to the coal type. The carbon content and particle size distribution of the unburned coal particles are analyzed by ASTM (American Standard Test Method) D5142-90(1994): *Standard Test Methods for Proximate Analysis of the Analysis Sample of Coal and Coke by Instrumental Procedures*. The burnt coal particles are sampled by US-EPA Method 5: *Determination of Particulate Matter (PM) emissions from Stationary Sources*, using the sampling configuration shown in figure 20.

Slagging

Slagging is the term for the phenomenon of the accumulation of ash on the walls of the combustion chamber. The tendency of a coal to cause ash deposition is affected by the components of the coal ash, as well as the temperature of the ash particles as they come in contact with a surface. Slagging will change the heat transfer properties of the walls of the combustion chamber. These changes are monitored by the slagging probe for the different coals tested in the test furnace. The slagging probe has the same dimensions as the temperature/gas probe shown in figure 17 except that the part placed in the furnace is closed. A thermocouple was welded on the top of the probe to measure the changes in temperature. The temperature differential between the outer and inner parts of the probe is translated to heat flux. Measurements are made at l/D=7.83 for 6-8 hours. A plot depicting the time vs. the decay of heat transfer as deposition occurs is drawn. The results for different coals are compared qualitatively only. It should be noted that GLACIER does not have the capability to predict slagging behavior.

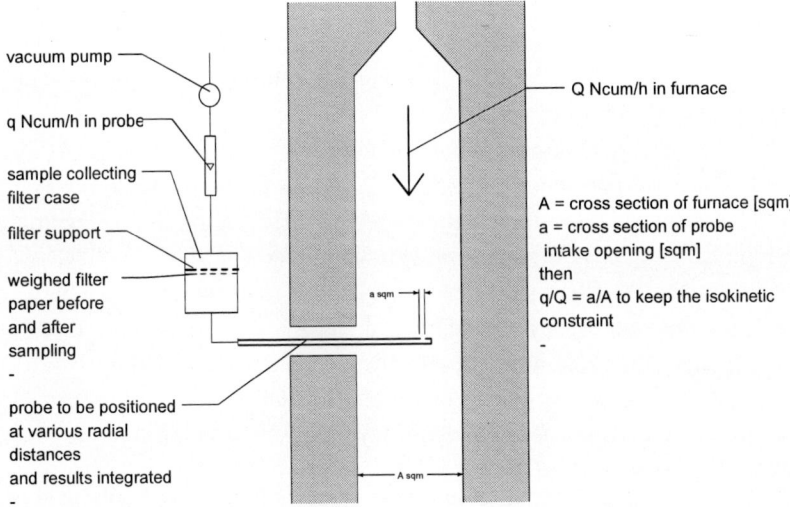

Figure 20. Isokinetic LOI sampling probe.

Coal/Char Nitrogen

For the GLACIER prediction of NO_x formation, the distribution of elemental nitrogen (N) between the volatile matter and the char needs to be known. Total nitrogen concentration in the coal and the nitrogen concentration in the coal char were analyzed using ASTM D3179-89 (1993): *Standard Test Methods for Nitrogen in the Analysis Sample of Coal and Coke*. Coal char samples were prepared by heating a 5 g sample of pulverized coal to 950 °C at a rate of 3 °C min^{-1}, kept at peak temperature for 6 min and then cooled at a rate of 3 °C min^{-1}.

2.3. Artificial Neural Network

The whole fitting process required many simulation runs, 25-30, that were carried out with discrete values of the model parameters. Since each simulation took 1-2 days until convergence, we developed an artificial neural network (ANN) [29-31] to reduce computational time and improve efficiency.

For the past 20 years, the potential of ANN as a universal modeling tool has been widely recognized. ANNs can manage complex and non-linear problems, processing information rapidly and thus reducing the engineering effort required in model development. Since ANNs approximate random non-linear functions and they do not need previous knowledge of the process phenomena, ANNs offer an alternative approach to model process behavior. They learn by extracting patterns from data that portray the relationship between inputs and outputs in any given process phenomenon. Neural networks use a trial and error method of learning. Training a neural network involves using a database of examples, which inputs and outputs are the values for the input and output of the network. An input is any information used to arrive at a solution, prediction or decision. An output is the solution, prediction or decision the ANN is being trained to generate. The neural network establishes the relationship between the inputs and the outputs by calculating their relative importance (weights). It calculates and compares its results to the actual example output. The network learns by adjusting the weights to minimize the output errors. In its basic form, a neural network is composed of several layers of neurons, an input layer, one or more hidden layers and an output layer. Each layer of neurons receives its input from the previous layer or from the network input. The output of each neuron feeds the next layer or the output of the network.

For this work, a feed-forward neural network has been trained to predict the kinetic model parameters of the coal combustion. The ANN was developed using MATLAB® and Simulink® (developed and marketed by The MathWorks, Inc.). The network was trained on the basis of modeling results for three coals: KPC (bituminous), Adaro (sub-bituminous) and KPC-Melawan (sub-bituminous). The input consists of 180 data sets, 60 for each coal. Each data set is comprised of 18 elements: 6 parameters (X – downstream distance from burner, T – gas temperature, NO_x concentration, O_2 concentration, coal index, and N – coal nitrogen content) for 3 different downstream (X) locations. The output data set also contains 18 elements. The neural network was trained using Levenberg–Marquardt optimization algorithm together with a cross-validation based on 'early stopping' mechanism to prevent over-fitting. During network training, the training algorithm continuously checks the network error on the test data set. The training was terminated at the point where the network error on the test data was a minimum. Early stopping is an implicit way to implement regularization, which can improve model robustness [29]. The training strategy implemented has the advantages of speed and not over-fitting the noise in the data.

2.4. GLACIER CFD CODE

For the last two decades, much effort was made to develop CFD models to simulate coal combustion in large-scale combustion chambers, such as utility boilers. Combustion system models aim to achieve two main objectives. One is operational problem solving, such as locating areas of corrosion in a combustion chamber. The second objective is pollutant emission reduction through operational strategies. Most of the models are capable of predicting good trend answers for the problems above. All of the models need to be compared to some measured values. However, once good agreement between the predicted and measured values is achieved, the model is used to give the full map of temperature and concentrations and other aspects in the furnace, which is not possible with any measurement system.

These models take into account fluid dynamics, local air/fuel mixing process, heat transfer and chemical kinetics. Coal combustion models are usually made up of four stages: (1) heating up, (2) devolatilization, (3) volatile combustion and (4) char combustion. Sub-models predict the pollutant formation, slagging and the physical aspects (heat transfer and flows). Much information about the combustion system is needed, such as combustion chamber geometry, feed, fuel and kinetic parameters. It is very important to use the correct coal volatilization parameters; otherwise there will be a large error in the model's results. Volatile combustion is treated with chemical reaction sub-models. Char combustion is more complicated and physical phenomena, such as boundary layer diffusion, needs to be taken into consideration [1].

Models were developed for single coal particle combustion as well as combustion systems and boilers. The single particle models contribute to the knowledge of the kinetic parameters and devolatilization and char oxidation rates, but they are unable to predict the behavior of large combustion systems, where the temperature and concentrations fields are not uniform. Therefore models of large scale systems were developed. These models concentrate on simulating the combustion and NO_x emissions, using CFD codes which are based on numerical finite volume solutions.

There are at least 15 different comprehensive combustion models reported in the literature [5]. Some of these are commercial; some are developed by the authors and most are a combination of a commercial code with sub-models developed by the authors. These models include various sub-models of the physical processes occurring in combustion systems, including gaseous fluid dynamics, radiative and convective heat transfer, homogeneous gas phase reactions, devolatilization, heterogeneous reactions, and particle motion. Most of these comprehensive combustion codes share common features, such as: (1) capability to model three-dimensional geometries; (2) k-ε two-equation turbulence model; (3) Lagrangian model for the entrained coal particles, with single- or two-step volatilization; (4) finite-volume discretization; and (5) SIMPLE or closely related scheme for the fluid dynamics solution. All the models share a common iterative solution scheme to achieve a converged solution.

As mentioned above, there are several commercial CFD codes on the market. For instance, the FLUENT code is popular for simulating combustion systems. As it is deployed in nearly every manufacturing industry, FLUENT is not specific enough for this research (simulating both test furnace and different types of utility boilers). Thus, it was decided to work with the GLACIER code, proprietary of Reaction Engineering International (REI) as it is specific for combustion systems. REI has extensive experience with combustion in utility boilers and has modeled over 150 different utility boilers firing a range of fuels including coal, oil, gas, biomass and blends of these fuels. Types of systems modeled include tangential-fired and wall-fired units, which are the type of furnaces modeled in this work. REI has expertise in pollutant formation, furnace performance and operational impacts.

Comparing the main modeling features of two commercial CFD codes, GLACIER and FLUENT, it was found that the computational features are similar except for three major differences. GLACIER uses only a rectangular elemental mesh while FLUENT enables both linear and curvilinear surfaces. This difference can become important when boundary layer effects along this surface are important to the development of the flow field. GLACIER uses a discrete-ordinate radiation model while FLUENT has a discrete transfer model for gaseous radiation problems. This difference can affect the outcome of chemical species, such as CO_2 and O_2. Lastly, although the NO_x chemistry is similar, the input parameters for the NO_x postprocessor sub-models are different.

The features of the GLACIER code that assure its predictive power are that the GLACIER code includes mass, momentum and energy coupling between the gas and particles and coupling between turbulent fluid flow, chemical reactions, radiative and convective heat transfer, and finite-rate NO formation. It assumes that the flow field is a continuum field that can be described locally by general conservation equations for mass, momentum, energy, and species. The computational approach involves numerical discretization of the partial differential equation set. Typically, 10^5-10^6 discrete computational nodes are used to resolve the most relevant features of a three-dimensional combustion process and approximately forty variables (including gas velocities, thermodynamic properties, and concentrations of various chemical species) are tracked at each node. Coal reactions are characterized by liquid vaporization, coal devolatilization, and char oxidation. Kinetic data required for running GLACIER are: nitrogen distribution between volatile matter and coal-char, six parameters for the two-step devolatilization mechanism ($Y_{1,2}$, $A_{1,2}$, $E_{1,2}$), three parameters for the one-step coal oxidation process (n, A_c, E_c), a set of parameters (plugged-in the code) for a comprehensive gas-phase NO_x mechanism starting from an inputted initial HCN-NH_3 ratio (ZEDA), one parameter for the conversion of char-nitrogen into NO; a total of twelve parameters, that none of them could be found in the literature. All of these parameters which describe the coal particles' combustion and NO_x formation rates can be varied in the GLACIER program to obtain the best agreement with the experimental data. After validating the GLACIER model for the test furnace, the same model parameters for the coal were used to model the utility boilers. The model validation using a test furnace before attempting prediction in a full-scale boiler strongly supports the full-scale predictions.

The approach described here determines a simplified kinetics model. While it cannot be said that the kinetic parameters used in the model will correspond to the combustion rate of a single particle of coal, these parameters do describe the combustion behavior of a "macroscopic" sample of the tested coal. For the goal of predicting the combustion behavior of these coals and blends in a utility boiler furnace, this simplified model used with the GLACIER code gave good results.

This chapter describes the GLACIER code and the different Configured Fireside Simulators (CFS), which were developed and purchased for the modeling tool.

GLACIER – Code Description

GLACIER, REI's proprietary reacting CFD software package was developed over the last three decades by researchers at the University of Utah, Brigham Young University, and REI. GLACIER is widely used to model the physical and chemical processes occurring in utility boilers. Simulating coal combustion and pollutant formation is stressed in the model code [8], which accounts for radiant and convective heat transfer, turbulent two-phase mixing, devolatilization and heterogeneous coal particle reactions (char oxidation), equilibrium (CO_2, O_2, H_2O, SO_x, CO) and finite rate (NO_x) gas-phase chemical reactions [32].

The GLACIER code includes mass, momentum and energy coupling between the gas and particles and coupling between turbulent fluid flow, chemical reactions, radiative and convective heat transfer, and finite-rate NO formation [33]. It assumes that the flow field is a continuum field that can be described locally by general conservation equations for mass, momentum, energy, and species. The computational approach involves numerical discretization of the partial differential equation set. Typically, 10^5-10^6 discrete computational nodes are used to resolve the most relevant features of a three-dimensional combustion process and approximately forty variables (including gas velocities, thermodynamic properties, and concentrations of various chemical species) are tracked at each node.

The Reynolds averaging of the conservation equations results in partial differential equations for the mean flow field variables. The equations are solved for the time averaged flow field variables. The comprehensive model uses an Eulerian-Lagrangian framework for modeling the gas and particle phases. The code couples the turbulent fluid mechanics and the chemical reaction process, using progress variables to track the turbulent mixing process and thermodynamic equilibrium to describe the chemical reactions associated with the main heat release chemistry.

GLACIER uses a Lagrangian model for particle conservation equations to predict the transport of the entrained particle phase. Particle properties such as burnout, velocities, temperatures, and particle component compositions are obtained by integrating the governing equations along the trajectories. Both gas and particle conservation equations include source terms for addition and loss of mass, momentum and energy due to interactions between the two phases. A coal particle is defined as a combination of coal, char, ash and moisture. Coal reactions are characterized by liquid vaporization, coal devolatilization, and char oxidation. The off-gas from particle reactions is assumed to be of constant elemental composition. Turbulent fluctuations and complete, local, complex chemical equilibrium are included in the particle reactions. Heat, mass and momentum transport effects are included for each particle.

The rate at which the primary combustion reactions occur is assumed to be limited by the rate of mixing between the fuel and the oxidizer, which is a reasonable assumption for the chemical reactions governing heat release. The thermodynamic state at each spatial location is a function of the enthalpy and the degree of mixing of two mixture fractions, one of which corresponds to the coal off-gas. The effect of turbulence and mixing on the mean chemical composition is incorporated by assuming that the mixture fractions are defined by a "clipped Gaussian" probability density function (pdf) having a spatially varying mean and variance. The mean and variance are computed numerically at each grid point and mean chemical species concentrations are obtained by convolution over the pdf. Specie concentrations are calculated as properties based on the local stream mixture and enthalpy. This is much more computationally efficient than tracking individual species. The radiative intensity field is solved based on properties of the surfaces and participating media, and the resulting local flux divergence (net radiant energy) appears as a source term in the gas-phase energy equation.

The numerical method is a based finite volume and the minimum running time for each case is a few days. The post processing is tabular (exit and LOI values) but mostly graphical using FIELDVIEW, by Intelligent Light, USA.

Most coal combustion modeling done today separates the models of nitrogen pollutants from the generalized combustion model. The models of nitrogen pollutants are executed after the flame structure has been predicted. The basis for this is that the formation of trace pollutant species does not affect the flame structure, which is governed by much faster fuel-oxidizer reactions. Another advantage of the approach is computational efficiency. The time required to solve the system of equations for the fuel combustion can require many hours of computer time while the pollutant sub-models typically converge in a small part (~10%) of the time required to converge the combustion case. Thus, NO_x sub-model parameters and pollutant formation mechanisms can be more easily investigated by solving the NO_x sub-model using a pre-calculated flame structure [16]. GLACIER cannot compute the full chemistry for all intermediate species, including nitrogen species. However, this detail is not required for engineering solutions for utility boiler combustion chambers. As long as all the physical mechanisms of first-order importance are also included in the investigation, engineering modifications to the boiler can be simulated using a condensed set of chemical kinetic mechanisms [8].

The Configured Fireside Simulator, CFS, is a Graphical User Interface, GUI, developed by REI for the purpose of running a preconfigured GLACIER model of a coal-fired furnace. CFS's for the test furnace and the opposite-wall and tangential-fired boilers were developed by REI for this project. The CFS's were validated by us by comparing different experimental and simulation results for each furnace type. To reach the stage where the methodology was fully developed over 100 simulations were run. Detailed descriptions of the different CFS's are described by Spitz Beigelman [34].

2.5. EXPERIMENTAL AND PREDICTIVE RESULTS

Coal Characteristics and Combustion Conditions

The elemental and proximate analyses of the investigated coals are presented in table 3. Table 4 details the combustion conditions for all coals and furnace types tested.

Table 3. Analyses of the tested coals

Parameter, %	Coal					
	SA	Ad	Dr	Gln	Ven	Mel
C	73.41	72.57	76.38	73.76	75.14	71.38
H	3.93	5.10	5.15	4.99	5.30	5.12
O	5.60	19.54	11.05	7.75	7.42	17.62
N	1.69	0.89	1.47	2.09	1.41	1.36
S	0.51	0.12	0.59	0.37	0.97	0.22
Ash	14.90	1.80	5.35	11.04	9.73	4.3
Total Moisture	7.60	25.70	13.56	9.32	7.50	22.3
Residual Moisture	3.10	14.50	6.01	3.35	1.91	15.5
Volatiles	27.90	50.10	41.54	33.91	36.76	47.1
Fixed C	57.20	48.10	53.11	55.05	53.51	48.6
Gross C.V., AR (MJ kg^{-1})	26.2	21.7	26.7	26.8	29.4	22.4
Net C.V., (MJ kg^{-1})	25.3	20.2	25.4	N/A	27.2	21.5

Test Furnace Results with KPC-Melawan Coal

Results for the test furnace fired with Drummond, Adaro and Venezuelan coals are presented elsewhere [37,38]. Figure 21 shows results for the KPC-Melawan coal fired in the test furnace. The results shown are the experimental and simulation results for the test furnace achieved with the set of model parameters that provided the best agreement between the experimental data and the simulations. In general, better agreement is found for temperature than for the gases. The reasons for this are the physical mechanisms of temperature dispersion and gas formation. The change in temperature is more gradual compared to changes in gas concentrations resulting from the many different chemical reactions occurring in the furnace.

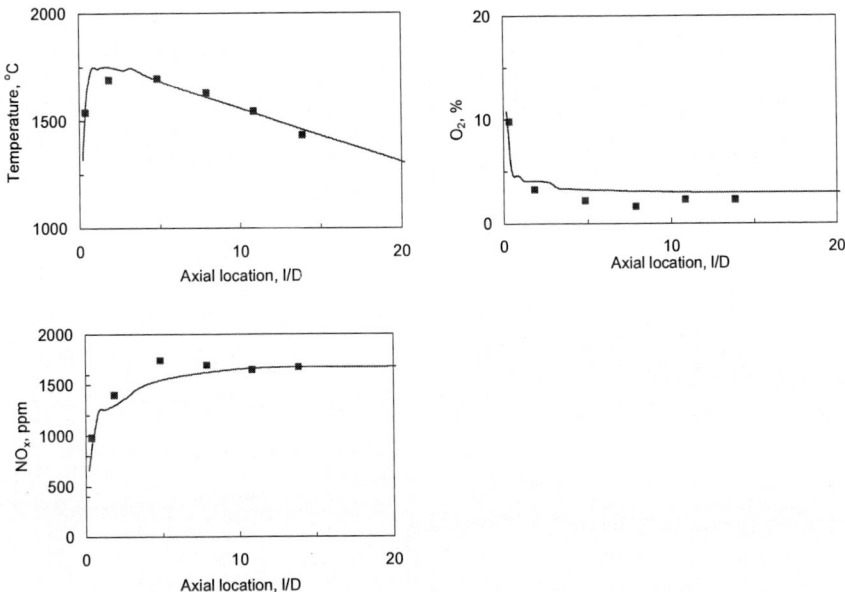

Figure 21. Experimental data (squares) and best agreement of numerical simulation (line) for temperature (top left), O_2 (top right) and NO_x (bottom left) for KPC-Melawan coal fired in test furnace.

Validation of the Model Parameters

The validity of the combustion kinetic model parameters was verified by comparison of different experimental data with simulation results obtained using the same set of combustion kinetic parameters. Good agreement between experimental data and simulation results for CO_2 and heat flux is shown in figure 22.

In addition, the agreement between simulation and measurement data is checked for different radial profiles in the test furnace. Figure 23 shows radial profiles at axial locations: l/D=1.83, representative of the near-burner zone, and l/D=13.83, representative of the plug-flow zone. The radial results are shown from furnace center to wall (from d/D=0 to 0.5, where d is the radial coordinate). In some cases the results at the wall are somewhat deviated due to dilution with external air and these points are not included here. The agreement between experimental data and numerical results is very good in the plug flow zone (l/D=13.83) for both temperature and NO_x. However, for NO_x the agreement to the radial profiles at l/D=0.33, which is very close to the burner (0.066 m), was not as good as the general trend. As in other works [1,25], attempts made to simulate very near-burner zone results for coal combustion did not succeed very much. Various reasons might be for this discrepancy; for example, the mesh was not fine enough, turbulent models are not suitable to these highly turbulent zones, or limitations of the sampling probe in this highly variable zone. However, good agreement between simulation results and experimental data was attained further downstream in the test furnace.

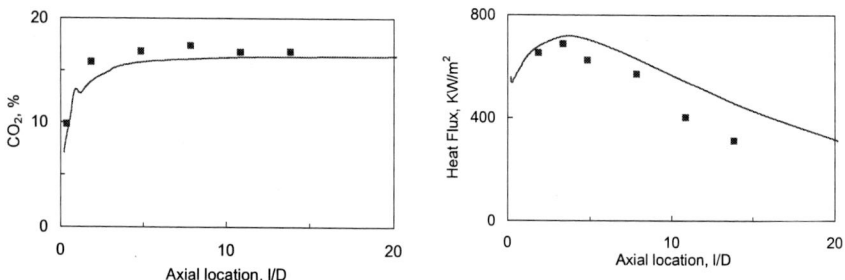

Figure 22. Experimental data (squares) and numerical simulation (line) results for CO_2 (left) and heat flux (right) for KPC-Melawan coal fired in the test furnace.

One must re-emphasize that the fitting procedure did not use radial profiles, but centerline profiles only. Thus, the good agreements for the radial profiles were an outcome of the whole process. Finally experiments were done at different operating conditions using the same model parameters to simulate the case with the new conditions. Example results for centerline NO_x profiles are given in figure 24 for KPC-Melawan coal fired in the test furnace with and without overfire air (OFA). For the OFA experiment a near-burner stoichiometric ratio of 0.85 was tested. OFA addition at l/D=13.83, illustrated in the dashed curve by the dip in NO_x concentration (caused by dilution), completed the total stoichiometric ratio to 1.15. Coal rate flow was the same as for the previous experiments. Good agreement between simulation results and experimental data was obtained. The model clearly predicts the influence of the near-burner lower stoichiometric ratio on the ~85% reduction of NO_x concentration at the test furnace exit.

Table 4. Combustion conditions for the tested coals

Parameter	Coal SA	Ad	Dr	Gln	Ven	Mel
Test furnace						
Primary air flow, Nm^3 h^{-1}	9.4	12.0	11.8	11.7	9.3	9.5
Primary air temperature, °C	67	60	60	60	59	64
Secondary air flow, Nm^3 h^{-1}	23.5	23.5	25.5	24.7	24.0	26.0
Secondary air temperature, °C	250	250	250	250	250	300
Total air flow, Nm^3 h^{-1}	32.9	35.5	37.3	36.4	33.3	35.5
Total coal flow, kg h^{-1}	4.1	5.2	4.4	4.58	3.8	5.0
Total heat rate, MJ h^{-1}	113	127	127	119	120	135
550MW Opposite-Wall Unit						
Primary air flow, Nm^3 h^{-1}	307000	371000	290000	287000	282000	358000
Primary air temperature, °C	78	60	78	60	78	60
Secondary air flow, Nm^3 h^{-1}	831000	759000	901000	837000	789000	971000
Secondary air temperature, °C	322	327	311	327	322	330
Total air flow, Nm^3 h^{-1}	1675000	1458000	1536000	1474000	1469000	1538000
Total coal flow, kg h^{-1}	200000	237000	181000	181000	169000	187000
Total heat rate, MJ h^{-1}	5271000	6356000	4864000	4854000	4532000	5044000
575MW Tangential-Fired Unit						
Primary air flow, Nm^3 h^{-1}	313000	381000	310000	375000	302000	302000
Primary air temperature, °C	70	60	70	60	70	60
Secondary air flow, Nm^3 h^{-1}	976000	928000	970000	926000	1054000	1054000
Secondary air temperature, °C	315	318	315	307	315	330
Total air flow, Nm^3 h^{-1}	1560000	1549000	1551000	1541000	1550000	1527000
Total coal flow, kg h^{-1}	196000	247000	193000	192000	177000	177000
Total heat rate, MJ h^{-1}	5266000	6624000	5188000	5149000	4747000	3705000

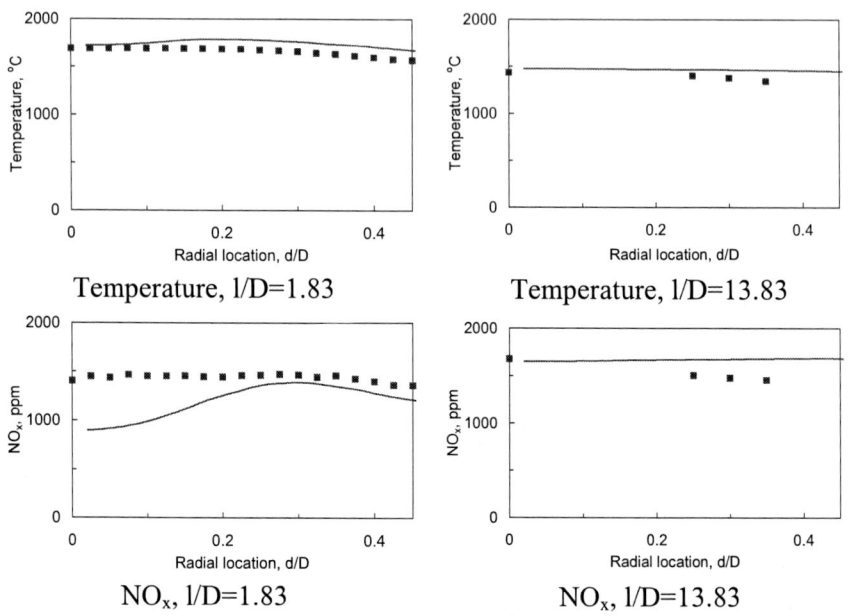

Figure 23. Experimental data (squares) and numerical simulation (line) of radial profiles at axial locations: l/D = 1.83 (left) and l/D = 13.83 (right), for temperature (top) and NO_x (bottom) when firing KPC-Melawan coal in test furnace.

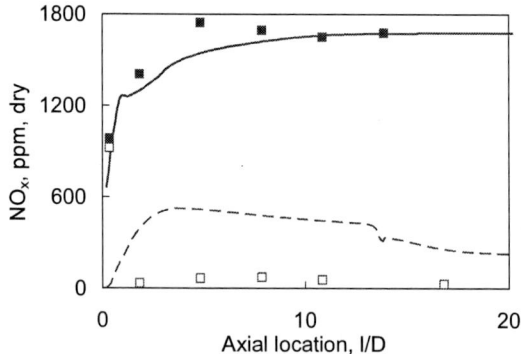

Figure 24. Experimental data (squares) and numerical simulation (line) results for NO_x from KPC-Melawan coal fired in test furnace at different firing conditions. Solid squares and line represent fired stoichiometric ratio of 1.15 (no OFA). Open squares and dashed line represent near-burner ratio of 0.85 and total stoichiometric ratio of 1.15 after OFA addition.

Table 5. Model parameters that provided best agreement between numerical results and experimental data for KPC-Melawan coal at all experimental conditions. Explanation on each parameter in text after table

Model Parameter	KPC-Melawan coal
Devolatilization	
Y_1	0.4
Y_2	0.8
A_1, sec^{-1}	375000
A_2, sec^{-1}	1.46E+13
E_1, kJ mol^{-1}	96.3
E_2, kJ mol^{-1}	251.0
Char oxidation	
n	1
A_c, kgC s^{-1} m^{-2} PaO$_2$$^{-1}$	6.55E-07
E_c, kJ mol^{-1}	87.9
NO$_x$ formation	
VMNFR	0.3
ZEDA	0.9
ZEDAH	0.1

The devolatilization reactions are defined by the following six parameters: Y_1 and Y_2 (the mass stoichiometry coefficients), A_1 and A_2 (devolatilization rate preexponential factors), E_1 and E_2 (devolatilization activation energies). The char oxidation rate and combustion behavior are defined by three global kinetic parameters: A_c (preexponential factor), E_c (activation energy) and n (reaction order). The parameters, VMNFR, ZEDAH and ZEDA, are used for the NO$_x$ postprocessor simulation. VMNFR defines the nitrogen division between volatile matter and char. ZEDA is the parameter that partitions the formation of volatile matter nitrogen between HCN and NH$_3$. ZEDAH specifies the fraction of char nitrogen that is converted to NO$_x$.

After verifying that the set of kinetic combustion model parameters used in the CFD model of the test furnace show good agreement between numerical results and experimental data in the test furnace plug flow zone for both axial and radial profiles and for different combustion conditions, we use the same set of model parameters for the same CFD model configured for full-scale boilers. Table 5 lists the values of the model parameters for KPC-Melawan coal, as determined from the procedure detailed in the Methodology section.

Full-Scale Predictions

The extracted combustion kinetic parameters are used for simulations of full-scale boilers using the same CFD code. We verified the full-scale model predictions with results from a series of full-scale firings done with different coals fired by IEC: three coals for the opposite-wall boiler and four coals for the tangential-fired boiler. The 550 MW drum type radiant opposite-wall fired unit was designed by Babcok and Wilcox (B and W) and the 575 MW tangential-fired unit was designed by Combustion Engineering (CE). The opposite-wall boiler is equipped with low-NO_x burners. Detailed descriptions of the boilers and furnace performance are described elsewhere [35,36] and a short description is provided here.

Boilers design

The evaluations were performed for a 550 MW drum type radiant opposite fired unit designed by Babcok and Wilcox (figure 25) and for a 575 MW unit equipped with tangentially fired boilers designed by Combustion Engineering (figure 26).

The 550 MW drum type radiant opposite wall unit (figure 25) comprises of two (2) balanced draft pulverized coal fired B and W drum type radiant boilers. These have a parallel backend arrangement for reheat steam temperature control, conservatively designed to supply steam to a single reheat steam turbine generator. Furnace walls utilize a gas-tight welded membrane construction. Internally, ribbed tubes are used in the furnace wall areas where the heat flux, fluid velocity and quality dictate their use to maintain nucleate boiling and minimum metal temperature. The furnace wall tubes are bent to accommodate the burners, observation ports, access doors and wall blowers. Integral windboxes are attached to the furnace walls of the unit in the burner zone for air distribution to the burners. Each opposite wall boiler is equipped with thirty (30) B and W DRB-XCL combination oil and coal burners. The burners are arranged in five rows high of six burners each elevation (3 burners are located at front wall and 3 burners at rear wall). The aerodynamic features of the DRB-XCL burner reduce oxygen availability during the early stages of combustion and redirect a portion of the secondary air further into the furnace to complete char burnout. The burner design also promotes a high temperature region just downstream of the nozzle exit to promote the conversion of fuel nitrogen to volatile molecular nitrogen and to thereby minimize the amount of fuel nitrogen retained in the char. Each opposite wall boiler is also supplied with six (6) B and W dual air zone over-fired air ports for reducing NO_x formation. The dual air zone NO_x port is designed to provide optimum mixing of air and flue gas in the second stage of combustion. NO_x port size selection is based upon full load operation with all the burners in service and a burner zone stoichiometry of 0.85. To improve and homogenize temperature distribution, an "aerodynamic nose" is installed in the furnace exit. The furnace bottom design is the open hopper type. The boiler consists of superheater and reheater elements. Steam capacity of the boiler is 1650 t/h, main steam/reheat steam pressures are 176/43 atm. Main steam/reheat steam temperatures are 540/540 °C.

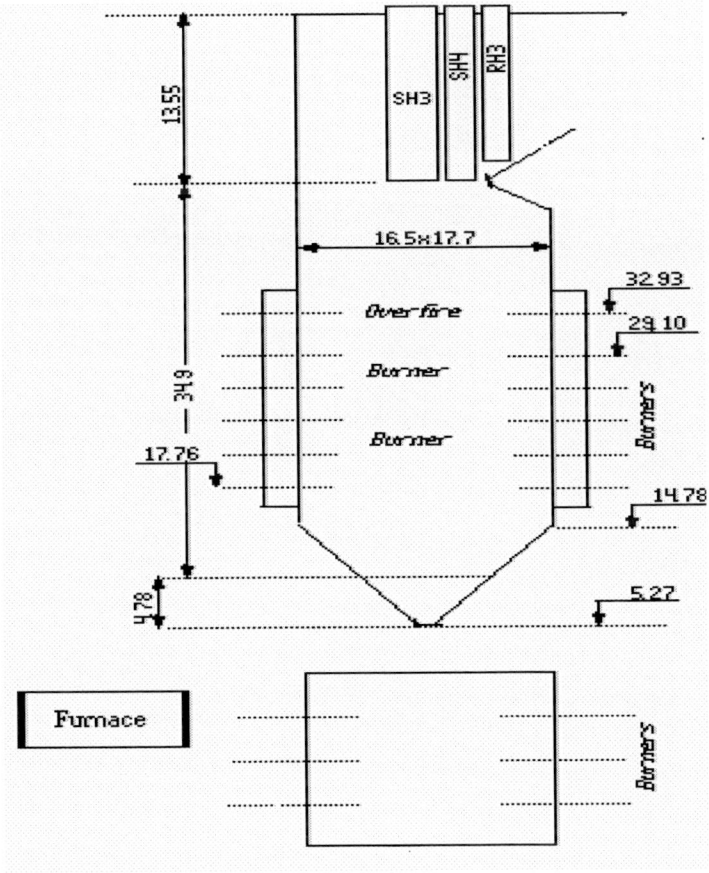

Figure 25. Scheme of B and W drum type radiant 550 MW opposite wall furnace.

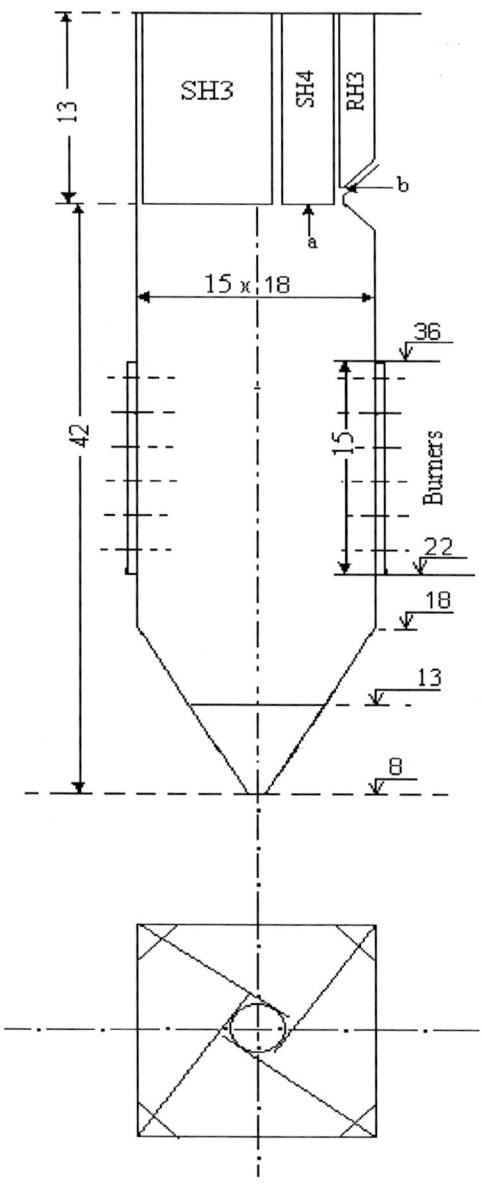

Figure 26. Scheme of CE 575 MW tangential fired furnace.

The CE 575 MW tangentially fired boiler (figure 26) is equipped with twenty straight flow burners, located in five levels, tangentially fired within the furnace. Five pulverizers, one for each level, are used. A fraction of the secondary air is fed through the closed-coupled overfire air ports located above the burners. Steam capacity of the boiler is 1700 t/h, main steam/reheat steam pressures are 181/43 atm. Main steam/reheat steam temperatures are 540/540 °C.

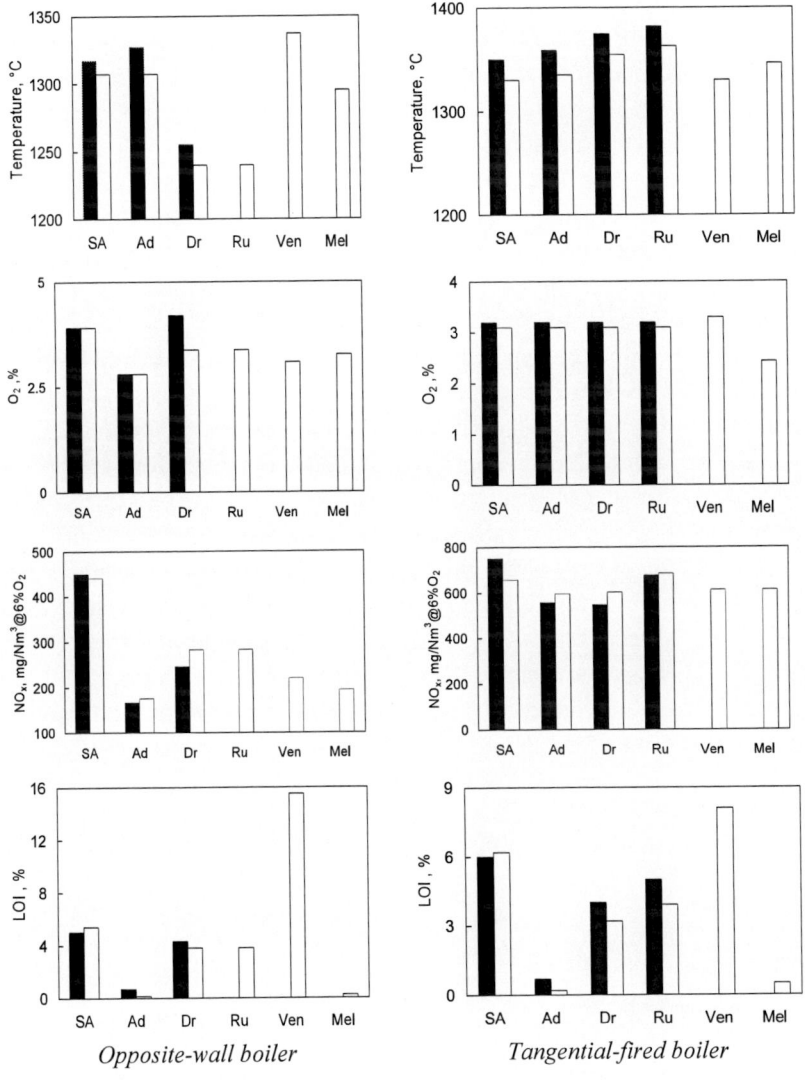

Figure 27. Mass-weighted averaged values before convective pass plane: full-scale tests (filled columns) and predictions (open columns) of temperature, O_2, NO_x and LOI for IEC 550MW opposite-wall boiler (left) and 575MW tangential-fired boiler (right).

Prediction Results

Figure 27 displays the results for both boiler types and compares experimental (filled columns) with simulation (open columns) results [37,38]. The results shown are the mass-weighted averages measured or calculated at the furnace exit, before the convective pass, for temperature, O_2, NO_x and LOI. Good agreement between experiment and simulation was obtained for the temperature. It should be mentioned that the temperature results obtained from IEC are not directly measured but are calculated by the heat balance code developed at IEC [39]. For both boilers, the results are about 10-25°C higher than the predictions. This gives an error of about 1-2% for the temperature, which is in the operational limits. Oxygen test results are very close to our predictions. In most of the cases the concentration differences are between 0-0.1%, which is an error of about 3%. An exception is the oxygen concentration for Drummond coal in the opposite-wall boiler, where the measured level is 25% higher than the predicted level. We do not have an explanation for the large difference compared to the other coals and boiler. When comparing the general NO_x levels for both boilers, the opposite-wall NO_x concentrations are about 45% less than those for the tangential-fired furnace fired with the same coals. It is apparent that the opposite-wall boiler is fitted with low-NO_x burners. The simulation results clearly show this reduction trend, with an average error of about 7-8%. There is an average error of about 32% in LOI results for both boilers. It is known that the models for char burnout need to be further developed [1]. However, because the general trend of measured LOI for the different coals is reproduced by the numerical simulations; extreme values determined by simulations should be researched further before actually firing the coal.

Chapter 3

PREDICTION RESULTS FOR COALS NOT TESTED BY IEC

3.1. COMBUSTION BEHAVIOR AND POLLUTANT EMISSION

As described in the previous chapter, the full-scale model predictions were verified with results from a series of full-scale tests done with different coals fired by IEC [37,38]. We verified our methodology for different coals well known to IEC: Billiton-BB Prime – a bituminous coal from South Africa, Glencore-Adaro – a sub-bituminous coal from Indonesia, Drummond-La Loma – a bituminous coal from Colombia and, for tangential-fired boiler only, Glencore-Russian – a bituminous coal from Russia. For both boilers, see figure 27, we predicted the behaviour and emissions from two coals previously unknown to IEC: Guasare-Venezuelan – a bituminous coal from Venezuela and KPC-Melawan – a sub-bituminous coal from Indonesia. For opposite-wall boiler we also simulated the combustion of Glencore-Russian coal. The combustion model parameters used in these predictions were obtained and verified according to the methodology described. The predictions were done using the operation parameters in table 4 and model parameters in table 5.

Predictions for the Russian coal fired in the opposite-wall boiler gave similar results to the Drummond coal. Further support to the validity of the simulation is shown in figure 27; the tested Russian coal is similar to Drummond also in the tangential-fired boiler.

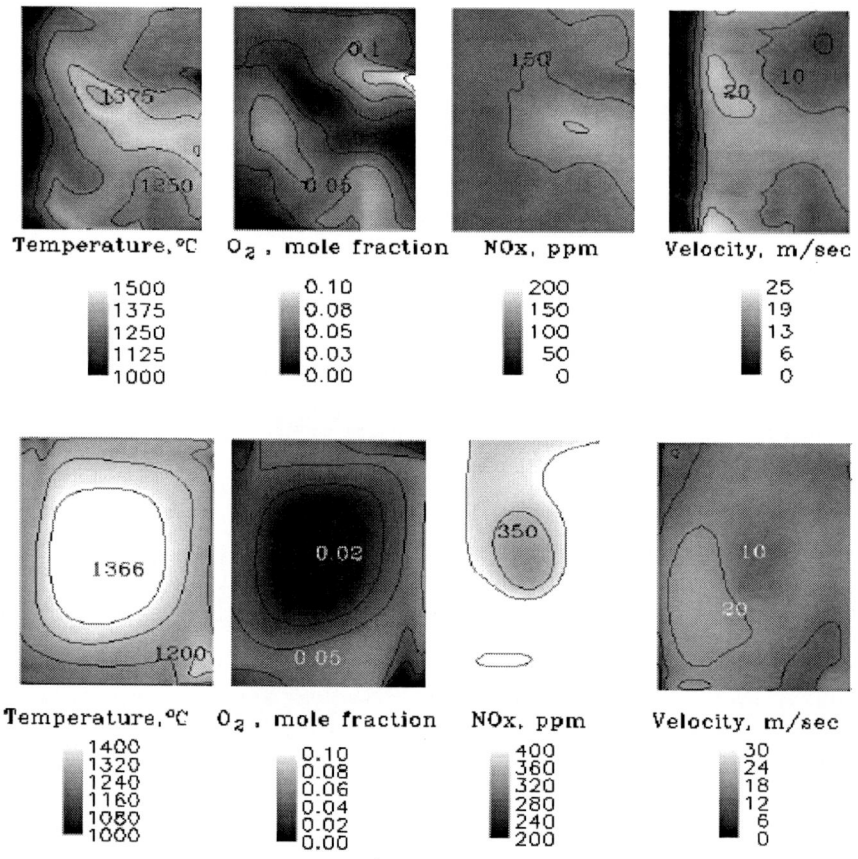

Figure 28. KPC-Melawan coal: predicted gas temperature, O_2 and NO_x concentrations and velocity at the plane before the convective pass of the opposite-wall boiler (top) and tangential-fired boiler (bottom).

The predicted mass-weighted averages at the furnace exit, before the convective pass, for temperature, O_2, NO_x and LOI are shown in figure 27 for the Venezuelan and Melawan coals for both boiler types. The Venezuelan coal shows a higher exit temperature for the opposite-wall boiler and a lower exit temperature for the tangential-fired boiler while KPC-Melawan coal has similar temperature values compared to the other coals. Still, the predicted temperatures meet the design temperatures for both boilers. The oxygen levels for both coals are similar to the others with the exception of KPC-Melawan fired in the tangential-fired boiler which shows a lower than average value. NO_x levels are similar to the average for both coals in both boilers. The KPC-Melawan coal shows similar behavior to Adaro, the other sub-bituminous Indonesian coal tested, in the lower than average NO$_x$ levels in the opposite-wall boiler. It is interesting to note for both boilers: the extremely high LOI levels predicted for Guasare-Venezuelan coal and the very low LOI levels predicted for KPC-Melawan coal, similar to Adaro. Figure 28 shows a comprehensive picture of the temperature, O$_2$, NO$_x$ and velocity at the exit plane (before the convective pass) of each boiler type, which is the area of most significance for the utility operator. The different parameters need to meet the design values for good operation or environmental requirements.

Table 6 presents predicted mass-weighted exit values for temperature, O$_2$, CO, NO$_x$ and LOI for both boilers for three different firing configurations for KPC-Melawan coal. The results of different firing configurations were received by varying the staged combustion stoichiometric ratios. This gives IEC a better representation of the influence of the unknown coal on combustion behavior and emissions. All parameters are within normal operational limits. The LOI results are low for full-scale boiler operation but since these results are comparable with the tested Adaro coal, we feel that the general trend of low LOI values is valid.

Table 6. Predicted mass-weighted averaged results for different firing conditions of IEC boilers fired with KPC-Melawan coal

	SR = Total stoichiometric ratio and (near-burner stoichiometric ratio)		
	550MW Opposite-wall furnace		
Parameter	*1.16(0.82)*	*1.16(0.86)*	*1.16(0.88)*
Temp., °C	1292	1295	1301
O$_2$, %	3.22	3.26	3.26
CO, ppm	20	27	30
NO$_x$, mg/dNm^3@6%O$_2$	174	190	193
LOI, %	0.31	0.83	0.23
	575MW Tangential-fired furnace		
Parameter	*1.2(1.0)*	*1.187(1.0)*	*1.187(0.98)*
Temp., °C	1314	1346	1361
O$_2$, %	2.56	2.84	3.13
CO, ppm	750	726	557
NO$_x$, mg/dNm^3@6%O$_2$	595	612	600
LOI, %	0	0.5	0

3.2. OPERATIONAL PARAMETERS USING EXPERT SYSTEM

Expert System - Description

IEC's boilers are equipped with an on-line supervision system called EXPERT SYSTEM. The purpose of the supervision system is to quantify the performance of the combustion and heat transfer processes in real time, reporting continuously on the controlled parameter deviations from their reference values. For prediction purposes the supervision system is used in off-line mode as "what-if-then" mode. EXPERT SYSTEM is programmed to give the boiler performance based on the coal's characteristics and boiler data and now it uses data provided by BGU. The EXPERT SYSTEM analyzes the BGU data and outputs information in a manner known to IEC. Figures 29a-32c describe the process of obtaining the coal's kinetic parameters by BGU and the input of this data to EXPERT SYSTEM by IEC.

The on-line supervision system used by IEC, EXPERT SYSTEM, is described in detail by Chudnovsky et al. [35]. The basic functional aim of the supervision system is to quantify the performance of the process (of all units and its elements) in real time, reporting continuously on the controlled parameter deviations from their reference values. For prediction purposes the developed supervision system is used in off-line mode as "what-if-then" mode. The system consists of a data acquisition module, data validation model, on-line interface system, calculation modules and a data storage module. The three most important independent calculation modules are:

1. Turbine and unit heat rate calculation.
2. Boiler performance and efficiency calculation.
3. Furnace performance calculation.

The first module is intended for turbine cycle performance calculation. The second module is based on an algorithm that enables the provision of on-line boiler efficiency, heat duty and cleanliness factors for each monitoring stage. The third module is based on the FURNACE Code, which can operate in the on-line and off-line modes. The FURNACE Code uses 3D-zonal calculation model of heat transfer, which is described by Karasina et al. [40]. The code calculates the distribution of the flue gases temperature, as well as absorbed and incident heat fluxes at the furnace walls. The furnace design, burners design and arrangement, radiant heat transfer properties of the flue gas and all the operating conditions are taken into account. Validation of the calculation results is done by comparison with full-scale furnace test data [41]. Besides data of flue gas temperature and flue heat flux distribution, the FURNACE code calculates the water-wall cleanliness and coal burn-out characteristics in the furnace and estimates the temperature of the water-walls, superheater and reheater tube metal. The slagging and fouling indices of the coal are calculated separately and then used for the FURNACE code module.

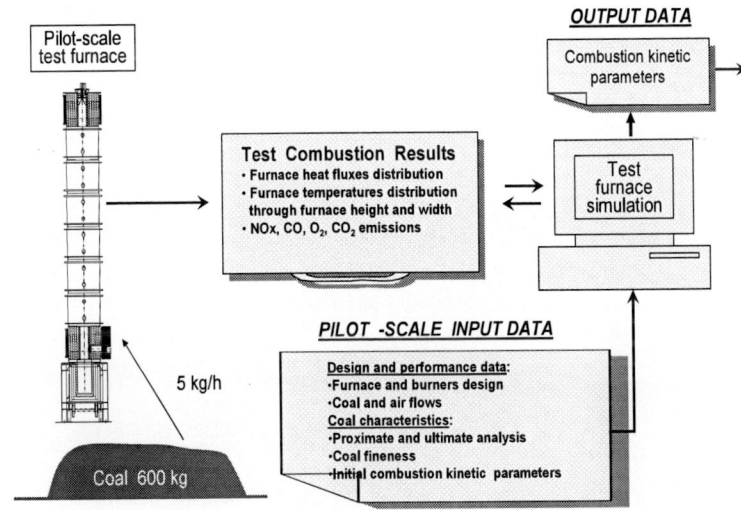

Figure 29a. Extraction of combustion kinetic parameters by BGU.

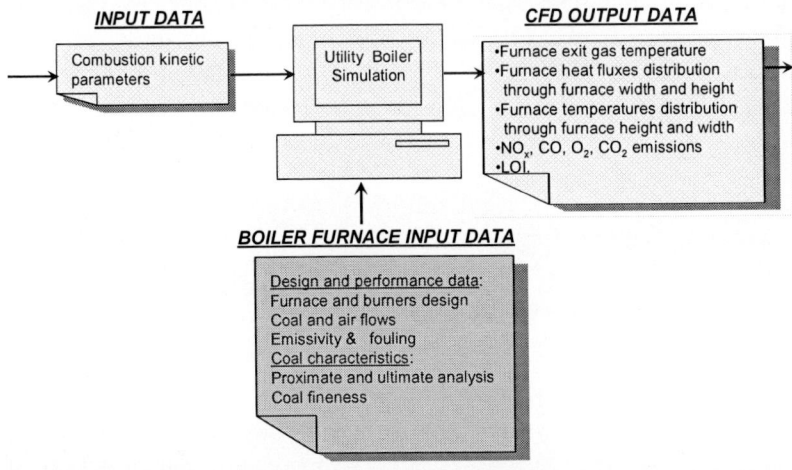

Figure 29b. Prediction of coal combustion behavior in boilers by BGU using boiler input information from IEC.

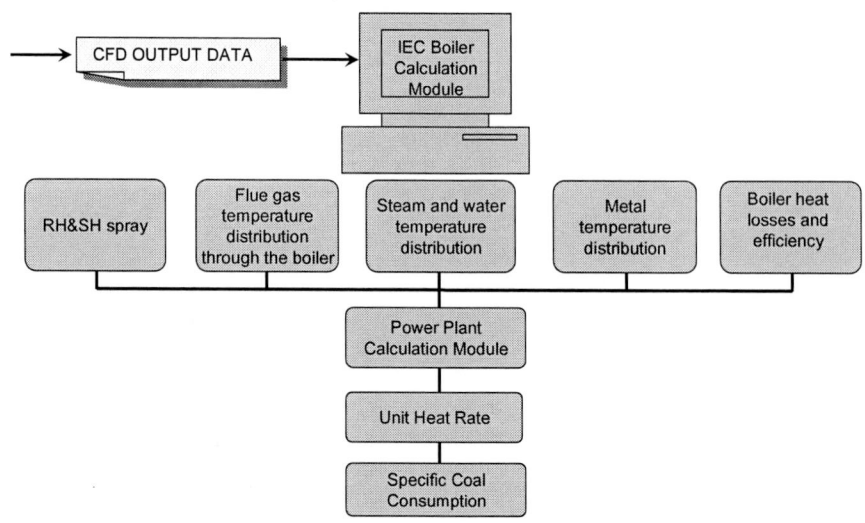

Figure 29c. Prediction and on-line supervision of coal combustion behavior in boilers by IEC using coal kinetic data given by BGU.

Prediction of Boilers' Performance Using Expert System

Chudnovsky and Talanker (2004) showed that fouling factor of the furnace is the function of the basic content of ash. Generalization of the obtained data for burning of certain coals [41] show that furnace cleanliness (fouling), under the same operation conditions, depends on ash characteristics. As was shown by Chudnovsky and Talanker (2004) the fouling factor of the furnace is a function of the basic content of ash, at NCR load for furnace clean condition after soot blowing. The basic content is equal to the ratio in Eq. 7:

$$Basic = \frac{B}{A+B} * 100\% \tag{7}$$

As follows, a basic content value increase leads to increase in fouling.

In order to clarify the influence of water wall absorptivity factor on radiation heat transfer process in the furnace, we provided a numerical analysis. For this purpose we ran a furnace calculation for different absorptivity factors, using commonly accepted water-wall fouling. Generalization of the obtained results enabled us to receive water-wall absorptivity as a function of acid ratio [41]. Approximation of the obtained results may be expressed by Eqs. 7 and 8:

For bituminous type ash:

$$a_w = -6.61\left(\frac{A}{A+B}\right)^2 + 8.05\left(\frac{A}{A+B}\right) - 1.19 \qquad (8)$$

$$\text{where } 0.82 < \frac{A}{A+B} < 0.95$$

For lignite type ash:

$$a_w = 1.81\left(\frac{A}{A+B}\right)^2 + 1.92\left(\frac{A}{A+B}\right) + 0.49 \qquad (9)$$

$$\text{where } 0.80 < \frac{A}{A+B} < 0.87$$

Due to low acid index of the ash the Melawan coal belongs to high emissivity ash and it will increase heat absorption in the furnace. From the other hand base index of the coal is very high and it may lead to increasing of fouling of the waterwall. However because of low ash content both factors influence should be less that in the case of typically burned, bituminous, coals.

Using the above data and the coal characteristics, we ran simulations of the opposite-wall and tangential-fired boilers for firing KPC-Melawan coal. The results are summarized in table 7.

Table 7. Predicted KPC-Melawan coal firing performance at NCR unit load

Parameter	Opposite wall boiler, 550 MW	Tangential fired boiler, 575 MW
Coal consumption, t/h	212	240.0
PA+FD Fan Air Flow, t/h	2040	2200.0
ID Fan Flue Gas Flow, t/h	2280.0	2460
SH spray, t/h	110-120	<5
RH spray, t/h	0.0	<3

Opposite-Wall Boiler

The total air flow is almost similar for KPC-Melawan coal in comparison with typical bituminous coals burning in IEC. At the same time, high water content in Melawan coal leads to a slightly increased flue gas flow (about 3%) and as a result a pressure drop increase about 7% is expected in flue gas path and therefore the ID Fan capacity is increased about 10%. This may cause a unit load limitation, especially for the summer period. Due to high water content in the burning coal, pulverizers inlet temperature is increased to approximately 240-260 °C. Pulverizers outlet temperature control is limited, however we expect to be able to provide pulverizers outlet temperature about 60 °C which is acceptable for stable ignition and firing. The primary air flow through pulverizes is about 95 t/h. Together with lower primary air temperature the ignition point will be located away from coal nozzles and will provide reliable operation of the nozzles. Superheater spray flow at full load is approximately 110 t/h and reheater spray flow is 0 t/h. Sulfur content in burning coal is 0.22% (DCB). Coal index (ratio of fixed carbon to volatile matter content) is equal to 1.064. Ash content reduction (4.3%) in comparison with typical coals decreases the unburned carbon losses and, as a result, boiler efficiency will increase when KPC-Melawan coal is fired. Evaluation is summarized in table 8.

Table 8. Predicted KPC-Melawan coal firing performance at NCR unit load

Parameter	Opposite wall boiler, 550 MW	Tangential fired boiler, 575 MW
Maximum mill capacity, t/h	52	52.5
Boiler heat losses, %	5.80	5.4
Unburned carbon heat losses, %	0.2	0.28
Efficiency, %	93.8	94.1
FEGT (preliminary)	1300	1370

Tangential-Fired Boiler

The total air flow for KPC-Melawan coal is slightly less in comparison with typical coals burning in IEC. At the same time, high water content in KPC-Melawan coal will lead to slightly increasing flue gas flow (about 3%) and as a result the pressure drop in flue gas path will increase about 7%. Therefore the ID Fan capacity will increase about 10% and will cause a unit load limitation, especially during summer period. This limitation may cause maximum achievable excess oxygen content at full load reduction up to 2.8%. This is less than boiler design data and may lead to increasing CO and unburned carbon in the ash. Due to high water content in the burning coal, pulverizers inlet temperature is increased to approximately 250-260 °C and in this case pulverizers outlet temperature control is limited. However a pulverizers outlet temperature about 60 °C can be provided which is acceptable for stable ignition and firing. The primary air flow through the pulverizers is about 95-98 t/h and together with lower primary air temperature the ignition point will move away from the coal nozzles and reliable operation of the nozzles can be achieved. Fuel air damper control is handled according to CE recommendation. The fuel oil and auxiliary air nozzle control compensates to keep the required windbox - furnace pressure drop. Superheater spray flow at full load does not exceed 5-10 t/h and reheater spray flow is equal to 0 t/h. Sulfur content in burning coal is 0.22% (DCB). Coal index (ratio of fixed carbon to volatile matter content) is equal to 1.064. Low ash content of 4.3% in comparison with typical, bituminous, coals will decrease the unburned carbon losses and, as a result, the boiler efficiency increases when KPC-Melawan coal is fired. Evaluation is summarized in table 9.

Table 9. Evaluation summary: IEC opposite-wall boiler behavior when firing KPC-Melawan coal

Topic	Evaluation
Coal type	Sub-bituminous coal
Ash type	KPC-Melawan coal belongs to lignitic ash category and corresponds to high slagging indices. Fe_2O_3/CaO indices are equal to 1.16 and may increase its fouling potential.
Ash softening temperature	The hemispheric ash temperature of KPC-Melawan coal is 1190° C. For lignitic ash category slagging index depends on AFT. For the current ash composition the coal has high slagging and low fouling potential. However, low ash content may prevent furnace slagging.
Mills capability	Due to low KPC-Melawan coal heating value, NCR load is provided by five pulverizers in operation
Mills outlet temperature	High moisture content in the coal requires raising primary air temperature before pulverizer in order to provide acceptable pulverizer exit temperature. Maximum possible pulverizers exit temperature 60 °C was achieved with primary inlet temperature about 240-260 °C.
PA, FD and ID Fans capability	Primary air flow through the pulverizers is about 95 t/h. PA+FD fan air flow is about 2040 t/h. ID Fan capacity is 2280 t/h. (This causes a unit load limitation, especially for the summer period.)
SO_2 and NO_x emission tendency	Low sulfur content in KPC-Melawan coal leads to reduced SO_2 concentrations. Fixed carbon to volatile matter ratio reduction also leads to lower NO_x formation.
Boiler performance	The boiler efficiency at NCR load is equal to 93.8% (low heating value base). SO_2 emission is equal to approximately 400 mg/dNm^3@6%O_2. NO_x emission is equal to 210 mg/dNm^3@6%O_2 at NCR load. LOI is equal to 4.0% at NCR load. Furnace exit temperature 1300 °C.
Ignition point	The selected pulverizer and firing system operation conditions provide reliable coal nozzle operation with acceptable distance of ignition point from the nozzle.

Table 9. Continued

Topic	Evaluation
Coal type	Sub-bituminous coal
Coal burnout	Coal burnout rate is slightly less in comparison with typical coal firing.
Furnace absorptivity	Furnace absorptivity is equal to 0.9-0.95 and fouling slightly less than for typical coal with the same ash composition.
Furnace performance	KPC-Melawan coal firing provides stable ignition and combustion process. Due low ash content incident heat flux less than for typical coal firing. Furnace exit temperature does not exceed allowable limit.
Metal temperature and steam temperature control	Superheater spray flow at full load was approximately 110 t/h and reheater spray flow was 0 t/h. Superheater spray provides steam temperature control in acceptable range. Water wall, SH and RH tube metal temperature less than allowable limit. Burner tilt and superheater spray provide steam temperature control in acceptable range. Flame position is the furnace is not symmetric and it is recommended to perform additional combustion tuning to correct this.

Table 10. Evaluation summary: IEC tangential-fired boiler behavior when firing KPC-Melawan coal

Topic	Evaluation
Coal type	Sub-bituminous coal
Ash type	KPC-Melawan coal belongs to lignitic ash category and corresponds to high slagging indices. Fe_2O_3/CaO indices are equal to 1.16 and may increase its fouling potential.
Ash softening temperature	The hemispheric ash temperature of KPC-Melawan coal is 1190° C. For lignitic ash category slagging index depends on AFT. For the current ash composition the coal has high slagging and low fouling potential. However, low ash content may prevent furnace slagging.
Mills capability	Due to low KPC-Melawan coal heating value, NCR load is provided by five pulverizers in operation
Mills outlet temperature	High moisture content in the coal requires raising primary air temperature before pulverizer in order to provide acceptable pulverizer exit temperature. Maximum possible pulverizers exit temperature 60 °C was achieved with primary inlet temperature about 250-260 °C.
PA, FD and ID Fans capability	Primary air flow through the pulverizers was about 95 t/h. PA+FD fan air flow about 2200 t/h. ID fan flue gas flow about 2460 t/h. Due to high moisture content in the Melawan coal the total flue gas flow is approximately 3% higher than typical coals fired which leads to an increase in the ID Fan capacity of about 9-10%. As result ID Fan achieved its allowable limit at 575 MW load and it causes unit load limitation especially during summer period.
SO_2 and NO_x emission tendency	Low sulfur content in KPC-Melawan coal leads to reduced SO_2 concentrations. Fixed carbon to volatile matter ratio reduction also leads to lower NO_x formation.

Table 10. Continued

Topic	Evaluation
Coal type	Sub-bituminous coal
Boiler performance	The boiler efficiency at NCR load is equal to 94.0% (low heating value base) with boiler exit temperature less than design value. SO_2 emission is equal to approximately 400 mg/dNm^3@6% O_2. NO_x emission is equal to 550 mg/dNm^3@6% O_2 at NCR load. LOI is equal to 2.5 % at NCR load. Furnace exit temperature 1370°C.
Ignition point	The selected pulverizer and firing system operation conditions provide reliable coal nozzle operation with acceptable distance of ignition point from the nozzle.
Coal burnout	Coal burnout rate is slightly less in comparison with typical coal firing.
Furnace absorptivity	Furnace absorptivity is equal to 0.9-0.95 and fouling slightly less than for typical coal with the same ash composition.
Furnace performance	KPC-Melawan coal firing provides stable ignition and combustion process. Due to low ash content, incident heat flux less than for typical coal firing. Furnace exit temperature does not exceed allowable limit.
Metal temperature and steam temperature control	Superheater spray flow at full load is less than 10 t/h and reheater spray flow 0 t/h. Burner tilt and superheater spray provide steam temperature control in acceptable range. SH and RH tube metal temperature less than allowable limit. Burner tilt and superheater spray provide steam temperature control in acceptable range.

Chapter 4

CONCLUSIONS

We successfully predicted performance and emissions of bituminous and sub-bituminous coal types from full-scale pulverized coal utility boilers of type: opposite wall and tangential fired. To predict the combustion behavior and pollutant emissions of coal in pulverized-coal utility boilers, we developed a methodology combining measurements in a 50kW pilot-scale test facility with simulations using the same CFD code configured for both test and full-scale furnaces. There is no attempt to predict the combustion behavior of the utility boiler based on the combustion behavior of the test furnace. For the goal of predicting combustion behavior of coals in a utility boiler furnace, our methodology gives good results. In addition to predicting combustion behavior and emissions, IEC developed an online supervision system called EXPERT system. This system calculates the different operator needed information.

REFERENCES

[1] A. Williams, R. Backreedy, R. Habib, J. M. Jones and M. Pourkashanian, *Fuel.* 81 (5), 605-618 (2002).

[2] A. Arenillas, R. I. Backreedy, J. M. Jones, J. J. Pis, M. Pourkashanian, F. Rubiera and A. Williams, *Fuel.* 81 (5), 627-636 (2002).

[3] A. M. Carpenter, *IEACR/81*. IEA Coal Research, London, 1995.

[4] J. M. Beer, Progress in Energy and Combustion Science. 26, 301-327 (2000).

[5] A. M. Eaton, L. D. Smoot, S. C. Hill and C. N. Eatough, *Progress in Energy and Combustion Science*. 25, 387-436 (1999).

[6] D. W. Pershing and J.O.L. Wendt, *Ind. Eng. Chem. Process Des. Dev.* 18(1), 60-67 (1979).

[7] H. Lorenz, E. Carrea, M. Tamura and J. Haas, *Fuel.* 79, 1161-1172 (2000).

[8] K-T Wu, H. T. Lee, C. I. Juch, H. P. Wan, H. S. Shim, B. R. Adams and S. L. Chen, *Fuel.* 83, 1991-2000 (2004).

[9] S. K. Ubhayakar, D. B. Stickler, C. W. Von Rosenberg and R. E. Gannon, *16th Symposium (International) on Combustion*. The Combustion Institute, 427-436 (1977).

[10] C. Sheng, B. Moghtaderi, R. Gupta and T. F. Wall, *Fuel.* 83, 1543-1552 (2004).

[11] R. I. Backreedy, R. Habib, J. M. Jones, M. Pourkashanian and A. Williams, *Fuel.* 78, 1745-1754 (1999).

[12] T. H. Fletcher and D. R. Hardesty, *SAND92-8209 UC-362*. Sandia Report (1992).

[13] J. M. Jones, P. M. Patterson, M. Pourkashanian, A. Williams, A. Arenillas, F. Rubiera and J. J. Pis, *Fuel.* 78, 1171-1179 (1999).

[14] R. H. Hurt and R. E. Mitchell, *24th Symposium (International) on Combustion*. The Combustion Institute, 1243-1250 (1992).

[15] D. Smoot, editor, "Fundamentals of coal combustion for clean and efficient use", *Elsevier Science Publishers B.V.*, The Netherlands (1993).
[16] S. C. Hill and L. D. Smoot, *Progress in Energy and Combustion Science*. 26, 417-458 (2000).
[17] P. Glarborg, A. D. Jensen and J. E. Johnsson, *Progress in Energy and Combustion Science*. 29, 89-113 (2003).
[18] A. Molina, A. G. Eddings, D. W. Pershing and A. F. Sarofim, *Progress in Energy and Combustion Science*. 26, 507-531, (2000).
[19] J. P. Spinti and D. W. Pershing, *Combustion and Flame*. 135, 299-313 (2003).
[20] A. Williams, M. Pourkashanian and J. M. Jones, *Progress in Energy and Combustion Science*. 27, 587-610 (2001).
[21] M. Rostam-Abadi, L. Khan, J. A. DeBarr, L. D. Smoot, G. J. Germane and C. N. Eatough, *American Chemical Society, Division Fuel Chemistry*. 41(3), 1132-1137 (1996).
[22] S. Schafer and B. Bonn, *Fuel*. 79, 1239-1246 (2000).
[23] J. A. Miller and C. T. Bowman, *Progress in Energy and Combustion Science*. 15, 287-338 (1989).
[24] P. J. Smith, S. C. Hill and L. D. Smoot, 19th *Symposium (International) on Combustion*. The Combustion Institute, 1263-1270 (1982).
[25] F. C. Lockwood, T. Mahmud and M. A. Yehia, *Fuel*. 77(12), 1329-1337 (1998).
[26] D. Gera, M. Mathur and M. Freeman, *Energy and Fuels*. 17, 794-795 (2003).
[27] R. Kurose, H. Makino and A. Suzuki, *Fuel*. 83, 693-703 (2004).
[28] J. C. P. Spinti, "An experimental study of the fate of char nitrogen in pulverized coal flames", PhD dissertation, The University of Utah (1997).
[29] Q. Zhu, J. M. Jones, A. Williams and K. M. Thomas, *Fuel*. 78, 1755-1762 (1999).
[30] J. Z. Chu, S. S. Shieh, S. S. Jang, C. I. Chien, H. P. Wan and H. H. Ko, *Fuel*. 82, 693-703 (2003).
[31] H. Zhou, K. Cen and J. Fan, *International Journal of Energy Research*. 29, 499-510 (2005).
[32] J. Valentine, M. Cremer, K. Davis, J. J. Letcavits and S. Vierstra, *2003 Power-Gen Conference*, December 9-11, 2003, Las Vegas, NV USA (2003).
[33] H. S. Shim, private correspondence (2004).

References

[34] N. Spitz Beigelman, Combustion of Coal Blends. PhD dissertation, Ben-Gurion University of the Negev, Israel (2006).

[35] B. Chudnovsky, L. Levin and A. Talanker, Advanced On-line Diagnostic for Improvement of Boiler Performance and Reduction of NO_x Emission. Proceedings of the PowerGen 2001 Conference, Brussels, Europe (CDROM).

[36] B. Chudnovsky, E. Karasina, B. Livshits and A. Talanker, Application of Zonal Combustion Model for On-line Furnace Analysis of 575 MW Tangentially Coal Firing Boilers. Proceedings of the PowerGen 1999 Conference, Frankfurt, Europe (CDROM).

[37] N. Spitz, E. Bar-Ziv, R. Saveliev, M. Perelman., E. Korytni, G. Dyganov and B. Chudnowsky, POWER2006-88065. Proceedings of ASME POWER2006, Atlanta, Georgia (2006).

[38] R. Saveliev, B. Chudnowsky, B. Kogan, E. Korytni, M. Perelman, Y. Sella, N. Spitz and E. Bar-Ziv, POWER2007-22065. Proceedings of ASME POWER2007, San Antonio, Texas (2007).

[39] A. Vikhansky, E. Bar-Ziv, B. Chudnovsky, A. Talanker, E. Eddings and A. Sarofim, *International Journal of Energy Research*. 28, 391-401 (2004).

[40] E. Karasina, Z. Shrago and S. Borevskaya, *Teploenergetika*. 7, 42-47 (1982) (in Russian).

[41] B. Chudnovsky and A. Talanker, Effect of Bituminous Coal Properties on Heat Transfer Characteristic in the Boiler Furnaces. 2004 ASME International Mechanical Engineering Conference, Anaheim, California, USA (CDROM).

INDEX

A

access, 44
acid, 54, 55
activation, 4, 6, 7, 8, 9, 14, 43
activation energy, 4, 7, 8, 9, 14, 43
algorithm, 4, 8, 32, 52
alternative, 32
amines, 11
ammonia, 10
aromatic rings, 11
ash, 6, 7, 30, 36, 54, 55, 56, 57, 58, 59
atmospheric pressure, 20
availability, 44
averaging, 36

B

behavior, v, 1, 2, 4, 14, 15, 25, 29, 31, 32, 33, 35, 43, 50, 51, 53, 54, 57, 58, 61
binary blends, 15
biomass, 34
blends, 14, 34, 35
boilers, v, 1, 2, 3, 4, 5, 15, 33, 34, 35, 36, 37, 43, 44, 48, 49, 50, 51, 52, 53, 54, 55, 61
burn, 6, 15, 53
burning, 5, 17, 54, 55, 56
burnout, 6, 14, 36, 44, 48, 58, 59

C

calibration, 26, 27, 30
California, 65
carbon, 6, 30, 56, 57, 58
cast, 20
catalyst, 28
category a, 57, 58
ceramic, 20, 26, 27
CFD, v, 1, 2, 3, 4, 5, 6, 9, 12, 13, 14, 15, 19, 20, 33, 34, 35, 43, 44, 61
char combustion, 4, 10, 33
chemical composition, 3, 37
chemical kinetics, 33
chemical reactions, 20, 27, 34, 36, 37, 39
cleaning, 27
CO_2, 22, 25, 26, 28, 34, 36, 39, 40
coal, v, 1, 2, 3, 4, 5, 6, 7, 8, 10, 11, 12, 13, 14, 15, 16, 17, 20, 25, 30, 31, 32, 33, 34, 35, 36, 37, 38, 39, 40, 41, 42, 43, 44, 48, 49, 50, 51, 53, 54, 55, 56, 57, 58, 59, 61, 64
coal particle, 6, 7, 8, 10, 14, 16, 30, 33, 34, 35, 36
codes, v, 2, 4, 6, 8, 33, 34
Colombia, 49
combustion, v, 1, 2, 3, 4, 5, 6, 8, 9, 10, 11, 12, 14, 15, 16, 20, 25, 27, 30, 32, 33, 34, 35, 36, 37, 38, 39, 40, 43, 44, 49, 51, 52, 53, 54, 58, 59, 61, 64
combustion chamber, 3, 6, 11, 12, 30, 33, 37

combustion characteristics, 14
combustion processes, 3, 15
compensation, 5
complexity, 14
components, 8, 26, 28, 30
composition, 1, 3, 6, 36, 57, 58, 59
compounds, 10, 11
concentration, 11, 15, 25, 29, 31, 32, 40, 48
condensation, 27
conduction, 25
conductivity, 29
confidence, 5
configuration, 4, 20, 22, 30
conservation, 35, 36
construction, 44
consumption, 55
control, 10, 15, 25, 44, 56, 57, 58, 59
convergence, 32
conversion, 10, 11, 14, 28, 35, 44
cooling, 27
correlation, 5, 26
corrosion, 33
couples, 36
coupling, 34, 36
credibility, 3
cross-validation, 32

D

data set, 32
database, 32
decay, 31
decision making, 1
definition, 6
dendrites, 17
density, 9
deposition, 30
destruction, 10
deviation, 28
devolatilization, 4, 6, 7, 8, 10, 11, 14, 33, 34, 35, 36, 43
diffusion, 9, 33
diffusivity, 9
discretization, 34, 35, 36
dispersion, 38

distribution, 4, 14, 16, 20, 30, 31, 35, 44, 52
divergence, 37
division, 13, 43
doors, 44
dosing, 16
draft, 44

E

electricity, 1
emission, 1, 33, 57, 58, 59
energy, 4, 8, 29, 34, 36, 37
energy transfer, 29
environment, 6, 10
environmental regulations, 1
EPA, 30
equilibrium, 8, 9, 36
equipment, 3, 25
Europe, 65
experimental condition, 43
expertise, 34
extraction, 22, 27

F

flame, 12, 14, 25, 37
flow field, 9, 34, 35, 36
fluctuations, 36
flue gas, 45, 52, 55, 56, 58
fluid, 33, 34, 36, 44
fouling, 53, 54, 55, 57, 58, 59
fuel, 1, 9, 10, 11, 12, 13, 15, 16, 17, 20, 33, 36, 37, 44, 57
fuel type, 10

G

gas phase, 9, 34
gases, 6, 7, 15, 22, 25, 26, 27, 38
Gaussian, 37
Georgia, 65
graph, 29

Index

H

heat, 3, 17, 20, 25, 29, 30, 33, 34, 35, 36, 37, 39, 40, 41, 44, 48, 52, 54, 55, 56, 58, 59
heat loss, 20, 56
heat release, 36, 37
heat transfer, 25, 30, 33, 34, 35, 36, 52, 54
heating, 8, 10, 13, 31, 33, 57, 58, 59
heating rate, 8, 10, 13
heterogeneity, 1
hydrocarbons, 6
hydrogen, 6, 10
hydrogen cyanide, 10
hydrolysis, 11

I

indices, 53, 57, 58
Indonesia, 49
industry, 34
infinite, 8
insulation, 20
intensity, 37
interactions, 36
interface, 52
interpretation, 4
isothermal, 6
isotherms, 21
Israel, v, 5, 65
iterative solution, 34

K

kinetic model, v, 1, 2, 4, 5, 32, 39
kinetic parameters, v, 1, 2, 3, 5, 7, 13, 14, 15, 22, 33, 35, 39, 43, 44, 52, 53
kinetics, 3, 6, 9, 14, 35

L

learning, 32
limitation, 56, 57, 58
linear function, 32
literature, 6, 8, 13, 14, 26, 34, 35
low temperatures, 8
LPG, 16, 20, 21

M

manufacturing, 34
market, 34
measurement, 25, 26, 27, 28, 29, 33, 39
measures, 28
media, 37
milligrams, 3
mixing, 8, 15, 17, 19, 20, 22, 24, 33, 36, 45
modeling, 6, 8, 14, 32, 34, 35, 36, 37
models, v, 1, 2, 12, 15, 33, 34, 37, 40, 48
modules, 52
moisture, 6, 36, 57, 58
moisture content, 57, 58
molecules, 28
molybdenum, 28
momentum, 34, 36
motion, 18, 34
movement, 20

N

Netherlands, 64
network, 32
neural network, v, 2, 4, 32
neurons, 32
New York, iii
nitric oxide, 10
nitrogen, 4, 6, 10, 11, 12, 13, 14, 31, 32, 35, 37, 43, 44, 64
nitrogen dioxide, 10
nitrogen oxides, 10, 11, 13
nitrous oxide, 10
nodes, 35, 36
noise, 33
numerical analysis, 54

O

obstruction, 26
oil, 16, 17, 34, 44, 57

Index

operator, v, 2, 3, 5, 30, 51, 61
optical properties, 3
optimization, 4, 32
oxidation, 6, 7, 8, 9, 10, 11, 13, 14, 33, 35, 36, 43
oxidation rate, 9, 33, 43
oxygen, 6, 9, 10, 11, 14, 20, 22, 29, 44, 48, 50, 56

P

parameter, 4, 5, 13, 14, 35, 43, 52
partial differential equations, 36
particle temperature, 8
particles, 6, 8, 12, 25, 27, 30, 34, 36
particulate matter, 15, 27
performance, 3, 5, 14, 34, 44, 52, 55, 56, 57, 58, 59, 61
permeation, 27
physical mechanisms, 12, 37, 38
pollutants, 12, 37
poor, 6
ports, 44, 47
power, 1, 34
power plants, 1
prediction, v, 2, 14, 31, 32, 35, 52
pressure, 9, 16, 55, 56
prevention, 27
probability, 9, 37
probability density function, 9, 37
probe, 26, 27, 29, 30, 31, 40
problem solving, 33
production, 1, 7
program, 6, 14, 35
promote, 44
pulse, 26

R

radiation, 25, 28, 34, 54
radius, 8
range, 14, 27, 28, 34, 58, 59
reaction mechanism, 11
reaction order, 10, 43
reaction rate, 6
reaction time, 8
reaction zone, 4
reactivity, 9, 10
real time, 52
reduction, 12, 33, 40, 48, 56, 57, 58
refractory, 26
relationship, 32
retention, 27
rings, 11
robustness, 33
Russia, 49

S

safety, 20
sample, v, 2, 4, 14, 26, 28, 31, 35
sampling, 20, 21, 26, 27, 30, 31, 40
sensitivity, 14
sensors, 15
series, 3, 4, 44, 49
silica, 26
simulation, v, 2, 3, 5, 12, 20, 32, 37, 38, 39, 40, 42, 43, 48, 49
software, 35
South Africa, 2, 49
spatial location, 37
species, 9, 10, 11, 12, 14, 27, 34, 35, 36, 37
speed, 20, 26, 33
stability, 25
stages, 33, 44
steel, 17
stoichiometry, 7, 13, 43, 45
storage, 52
strategies, 1, 33
sulfur, 6, 57, 58
summer, 56, 57, 58
supervision, v, 2, 5, 52, 54, 61
supply, 44
surface area, 9, 10
swelling, 6
symmetry, 4
systems, 9, 10, 11, 25, 27, 33, 34

Index

T

temperature, 7, 8, 9, 10, 11, 13, 14, 15, 16, 20, 21, 22, 24, 25, 26, 27, 29, 30, 31, 32, 33, 38, 39, 40, 41, 42, 44, 47, 48, 50, 51, 52, 56, 57, 58, 59
temperature gradient, 29
test data, 33, 53
Texas, 65
thermal properties, 3
thermodynamic equilibrium, 36
thermodynamic properties, 35, 36
time, 7, 8, 10, 11, 12, 13, 20, 31, 32, 36, 37, 55, 56
time frame, 7
tracking, 37
training, 32
transformation, 30
transport, 36
trend, 33, 40, 48, 51
trial, 32
trial and error, 32
turbulence, 34, 37
turbulent mixing, 36

U

uncertainty, 5
uniform, 15, 19, 20, 22, 33

V

validation, 35, 52
validity, 9, 39, 49
values, 3, 7, 32, 33, 37, 43, 47, 48, 50, 51, 52
variability, 1
variable, 1, 8, 9, 40
variables, 12, 35, 36
variance, 37
vector, 19
velocity, 15, 19, 20, 22, 44, 50, 51
Venezuela, 49
volatilization, 33, 34

W

wavelengths, 28

Y

yield, 8, 13, 14